AF602043

Environmental Issues

The Author

Dr. Sudha Sambyal Malik is graduate in Basic Sciences from CSKHPKV Palampur, Himachal Pradesh and postgraduate in Environmental Sciences from the same institute. She awarded Ph.D degree in Environmental Sciences from Maharshi Dayanand University Rohtak, Haryana. She has qualified UGC - NET 2011 and 2012. She has published various research papers and review articles in National and International journals of repute and book chapters in the field of Organic Agriculture and Environmental Biotechnology.Presently she is working as Assistant Professor in Environmental Sciences in A.I.J.H.M. College Rohtak, Haryana.

Environmental Issues

Dr. Sudha Sambyal Malik

2020

Daya Publishing House®

A Division of

Astral International Pvt. Ltd.

New Delhi – 110 002

ISBN: 9789390371143 (Int. Edition)

Published by : Daya Publishing House®
A Division of
Astral International Pvt. Ltd.
– ISO 9001:2015 Certified Company –
4736/23, Ansari Road, Darya Ganj
New Delhi-110 002
Ph. 011-43549197, 23278134
E-mail: info@astralint.com
Website: www.astralint.com

Digitally Printed at : Replika Press Pvt. Ltd.

Preface

Environmental science is an interdisciplinary field. Because environmental disharmonies occur as a result of the interaction between humans and the natural world, we must include both when seeking solutions to environmental problems. In recent years, many environmental issues of global proportions have emerged, among them global warming and the destruction of ecosystems are the major one. The environment as an issue is ubiquitous in the world we live in. There has been a consequent explosion of work that reflects on the reasons for and responses to ecological dilemmas. The importance of environmental protection and conservation measures has been increasingly recognized during the past two decades. It is now generally accepted that economic development strategies must be compatible with environmental goals. This requires the incorporation of environmental dimensions into the process of development. It is important to make choices and decisions that will eventually promote sound development by understanding the environment functions. Therefore the present book entitled 'Environmental Issues' has been written so that essentials of Environmental issues could be made available to the readers in easy and readable manner.

The book contains all the topics included in environmental issues for the post graduate classes. The content of the book has been concentrated on familiarizing the student with basics of various environmental problems along with their solutions. The students that would be using this book come from exceedingly diverse backgrounds: some from medical background, some from non-medical and others from non-scientific background. For many, however, this book will the first contact with environmental issues and for many it would be a familiar book. Therefore, it has been tried to explain each and every topic in very simple and easy manner. In the present book, we have arranged the chapters to suit the requirement of the students and we are hoping that the new comers can find a useful introduction to environmental issues. Furthermore, the authors have endeavoured to present a balanced view of issues, diligently avoiding personal biases and fashionable philosophies.

Book is intended as a text for a one-semester, introductory course for post graduate students. They will find it interesting and informative. The central theme is interrelated-ness. No text of this nature can cover all issues in depth. What we have done is to identify major issues and give appropriate examples that illustrate the complex interactions that are characteristic of all environmental problems. There are many facts-presented in charts, graphs, and figures, that help to illustrate the scope of environmental issues. However, this is not the core of the text, since the facts will change. The book contains all the regional and global environmental issues. Regional issues include: deforestation, mining, siltation, eutrophication, wetland conservation, biodiversity conservation etc. Whereas, global issues include: greenhouse effect, ozone layer depeletion, global warming, air pollution, industrialization, urbanization, population growth etc. All these collectively will help to understand the importance of environment protection.

Dr. Sudha Sambyal Malik

Foreword

The human being is in search of new and sustaining relationships to the mother earth amidst environmental crises. While the particular causes and solutions of these crises are of great concern to the scientists, economists and policy-makers, the facts of widespread destruction are causing alarm in many quarters. Indeed from some perspectives the existence of human life itself appears threatened. The existence of human species has become questionable on an endangered earth planet. Pollution, solid waste management, conservation and quality of forests, growing water scarcity, falling of groundwater level, biodiversity depletion, soil degradation are the major environmental issues posing serious threat to the survival of organisms including mother earth itself. These issues need to be discussed, debated and deliberated among the students, intellectuals and environmentalists. One of sincere, brilliant and hard working students has taken a lead in this context by writing a book on "Environmental Issues".

It is my great pleasure to write a few lines about the book "Environmental Issues" written by Dr. Sudha Sambyal Malik Deptt. of Environment Sciences, A.I.J.H.M. College, Rohtak. We have many unions of opinions regarding the environmental issues in India. Dr. Sudha Sambyal is an eminent teacher and researcher in the field of Environmental Sciences. She has played, playing and will continue to play a key role to take up the environmental issues. This book is an outcome of her hard work and understanding of various environmental aspects and issues. The efforts put up by the author in writing this book deserve appreciation. I sincerely hope that this book will be very useful for the students, researchers and policy-makers working in the field of environmental sciences specifically dealing with environmental issues.

Dr. Ramesh. C. Chauhan

Senior Professor of Environmental Sciences

College of Basic Sciences

CSK, Himachal Pradesh Krishi Vishvavidyalaya

Palampur (Himachal Pradesh) - 176062

Contents

1

Ozone Layer Depletion

Environmental Issues

Our planet earth has a natural environment, known as 'Ecosystem' which includes all humans, plant life, mountains, glaciers, atmosphere, rocks, galaxy, massive oceans and seas. It also includes natural resources such as water, electric charge, fire, magnetism, air and climate.

Modern technologies used in the engineering and manufacturing industry have a major impact on our life in past few years. Due to the rapid changes in the engineering and manufacturing industries, drastic changes have been occurred in the environment.

Crucial environmental issues are no more a blame game. While most of us crib about dirty air, smelly garbage or polluted water, least do we know it is "us" who is responsible for this unfavorable circumstances leading to cautionary environmental issues. It is high time for human beings to take the 'right' action towards saving the earth from major environmental issues. If ignored today, these ill effects are sure to curb human existence in the near future. Flow chart showing various types of environmental issues is depicted in figure 1.1.

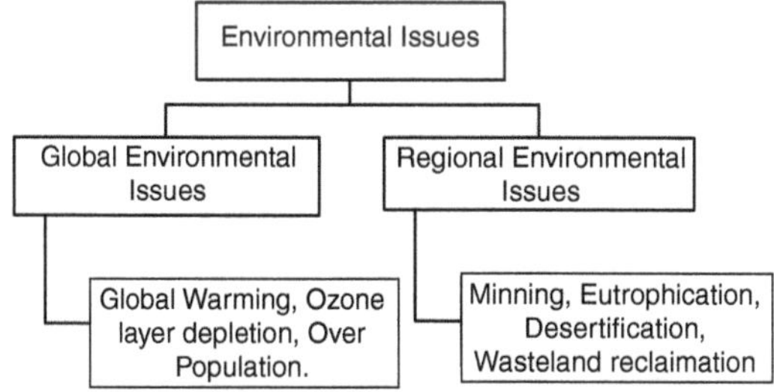

Figure 1.1 Flow chart showing various types of Environmental Issues

Ozone Layer Depletion

Ozone

Ozone is a tri-atomic molecule of oxygen. It is pale blue in color and also known as 'Good Up and bad nearby' because it is present both in troposphere and stratosphere but it act as pollutant in the troposphere and as a savior or protector in stratosphere. The ozone layer was discovered in 1913 by the French physicists Charles Fabry and Henry Buisson.

Ozone is present in the atmosphere as a layer known as ozonosphere, approximately 20-30 kms above the earth surface. Here it protects the earth by absorbing the harmful UV radiations of sun. UV radiations can be classified as UV-A, UV-B and UV-C. The UV-C radiations are the most dangerous but due to very short wavelength they do not Rach the earth surface. The UV-A radiations are not much harmful and UVB ra diations are very harmful, but these are absorbed by the ozone layer present in the earth's atmosphere.

Ozone Formation

Ozone is formed both in stratosphere and troposphere as given under:

In stratosphere it is formed naturally by a photochemical reaction as:

$$O_2 + hv\ (\text{light}) = [O] + [O]$$

$$O_2 + [O] = O_3$$

The ozone formation in stratosphere is in equilibrium with rest of the air.

In troposphere, it is formed as secondary pollutant as a result of a following photochemical reaction.

NOx (From industries) + VOC (Volatile organic compounds from industries) + hv (light) = O_3

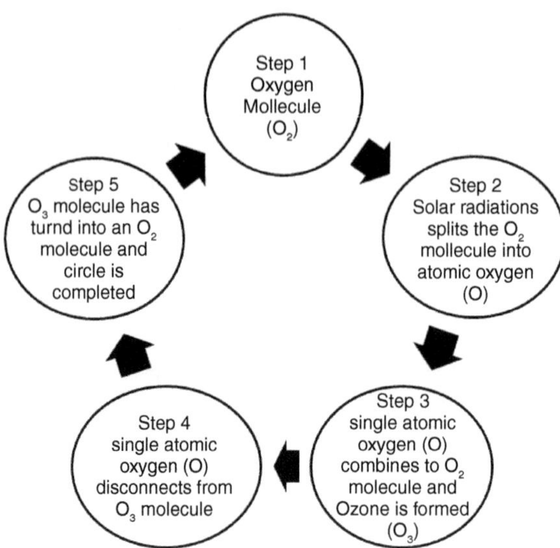

Figure 1.2 Oxygen-Ozone-Oxygen Cycle

Ozone Depletion (Ozone Hole)

Ozone hole refers to the thinning of the ozonosphere. It was first observed by three scientists, Joseph Farman, Jonanthan Shanklin and Brian Gardiner in 1985 over Antarctica. The thickness of ozone is measured by LIDAR (Light detection and ranging) and UV spectroscopy and it was first measured by G M B Dobson. Thus, the unit for ozone measurement is Dobson Unit (DU).

The actual thickness of ozone layer is 300 DU and when the thickness becomes less than 220 DU then it is called as ozone hole or ozone thinning. 1 DU is equivalent to 0.01 mm at standard temperature and pressure which is 273.5 Kelvin (0°C) and 1 atm pressure (760 mm) of Hg.

Process of Ozone Depletion

Anthropogenic activities- Ozone is depleted by anthropogenic activities as follows:

$$CFCl_3 + hv\ (\text{light}) = CFCl_2 + Cl^0 \quad (1)$$

$$Cl^0 + O_3 = O_2 + ClO \quad (2)$$

$$ClO + [O] = Cl^0 + O_2 \quad (3)$$

The reaction 2 & 3 are repeated again and again and it is estimated that one chlorine atom can destroy 1 00 000 ozone molecules.

Natural activity- Maximum ozone depletion occurs due to the natural activity of polar stratospheric clouds (PSCs) which are formed in winters over Antarctica. In winters the winds of Antarctica called the polar vortex, separates the region. Due to this the PSCs are formed. This act as chemical reactors and these contain ice, nitric acid and sulphuric acid. In these PSCs chemical reactions occurs and CFCs are formed and these destroy the ozone layer. The reaction occurs in the ice crystals and that is why in summer, when these ice crystals melt no such process is seen.

Cause of Ozone Layers Depletion

Chlorofluorocarbons (CFCs)

They are compounds formed by chlorine, fluorine and carbon. They are often used as refrigerants, solvents, and for the manufacture of spongy plastics. The most common are CFC-11, CFC-12, CFC-113, CFC-114, and CFC-115 which respectively have an ozone depletion potential of 1, 1, 0.8, 1, and 0.6.

Chlorofluocarbons, the chemicals used as the propellant for aerosol cans and Bromofluocarbons, Halon, are destroying the earth's Ozone layer. These chemicals were used in Freon and for fighting fires. Manufactures thought the chemicals were inert and not harmful to the environment.

When the chemicals reached the earth's stratosphere, they reacted with Ultraviolet radiation, which caused them to break down and release Chlorine and Bromine into the earth's ozone layer. The Ozone layer protects the earth from UV-B Rays. The chemicals caused a reaction, which made the ozone layer break down into pure oxygen. The layer lost its shielding effect from the suns, UV rays. The Bromine and Chlorine kept interacting with the ozone molecules until they eventually left the ozone layer to bond with other compounds. The contribution of various sectors in the emission of ozone depleting substances is given is figure 1.3.

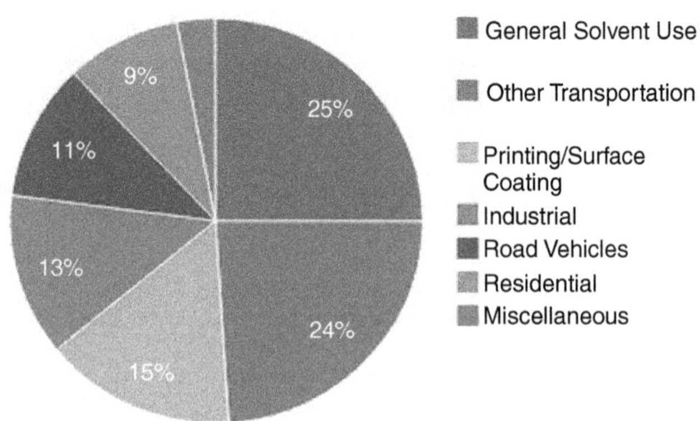

Figure 1.3. Contribution of various sectors in the emission of ozone depleting substances

Hydrochlorofluorocarbons (HCFCs)

These compounds are formed by H, Cl, F and C. They are being used as substitutes for CFCs because many of their properties are similar and are less harmful to ozone by having a shorter half-life and releasing fewer Cl atoms. Ozone depletion potential is between 0.01 and 0.1, but they remain harmful to the ozone layer. They are considered only a temporary solution and these have been banned in developed countries since the year 1930.

Halons

They are compounds formed by Br, F and C. Because of their ability to put out fires they are used in fire extinguishers. Although their manufacture and use is prohibited in many countries because of their ozone-depleting action. Their ability to harm the ozone layer is very high because they cantain Br which is a more effective atom destroying ozone than the Cl. Thus, halon 1301 and halon 1211 have ozone depletion potentials of 13 and 14 respectively.

Methyl bromide

It is very effective pesticide that is used to fumigate soils and in many crops. Due to its contents it damages the ozone layer and has an ozone depletion potential of 0.6. In many countries dates have been set around 2000, from which it will be banned.

Carbon tetra-chloride

It is a compound that has been widely used as a raw material in many industries, for example, to manufacture CFCs and as a solvent. It is also used as catalysts in certain processes where chlorine ions need to be released. It is no longer used as a solvent, as it was found to be carcinogenic. Its ozone depletion potential is 1.2.

Effects of Ozone Layers

Effects of the Depletion of the Ozone Layer on Human Health

Skin Cancer

Today, it is estimated that skin cancer rates increased due to the decrease in stratospheric ozone (ozone layer). The most common type of skin cancer, called non-melanoma, is the cause of exposures to UV-B radiation for several years. There are already people who have received the dose of UV-B that can cause this type of cancer.

The United Nations Environment Program (UNEP) predicts that at an annual rate of 10 per cent ozone loss over several decades, the increase in skin cancer will be around 250,000 per year. Even taking into account existing agreements for the phase-out of ozone-depleting substances (ODS), a realistic model would indicate that skin cancer would increase to 25 per cent above the level of 1980 by the year 2050, along of the 50 ° North latitude. The most lethal skin cancer, called melanoma, could also increase its frequency.

The Immune System

A person's defense against infection depends on the strength of his immune system. It is known that exposure to ultraviolet light reduces the effectiveness of the immune system, not only relating to infections to the skin but also to those that can be verified in other parts of the body.

Exposure to UV-B radiation may well enable the immune system to tolerate disease rather than combat it. This could mean the uselessness of vaccination programs in both industrialized and developing countries. Various effects of ozone layer depletion are given in figure 1.4.

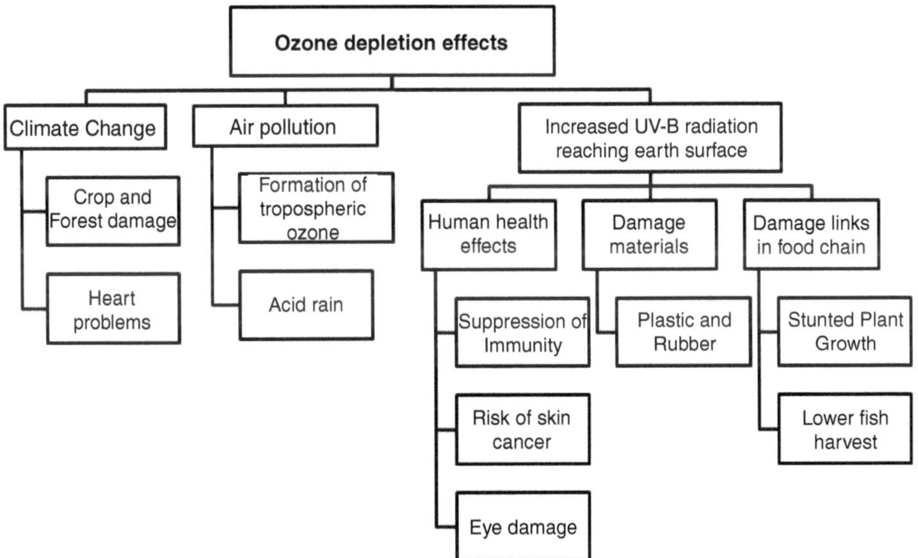

Figure 1.4. Environmental Effects of Ozone Layer Depeletion

Effects on Aquatic Ecosystems

The loss of phytoplankton, the basis of the marine food chain, has been observed as the cause of the increase in ultraviolet radiation. Under the ozone hole in the Antarctic phytoplankton productivity decreased between 6 and 12 per cent. It inhibits the reproductive cycle of phytoplankton, single-celled organisms such as algae that make up the bottom rung of the food chain.

UNEP indicates that a 16 per cent decrease in ozone could result in a 5 per cent loss of phytoplankton, which would mean a loss of 7 million tonnes of fish per year – around 7 per cent of global fish production. Thirty per cent of human consumption of protein comes from the sea, this proportion increases even more in the developing countries. Researchers also have documented changes in the reproductive rates of young fish, shrimp, and crabs as well as frogs and salamanders exposed to excess ultraviolet B.

Effects on Terrestrial Ecosystem

Animals

For some species, an increase in UV-B radiations implies the formation of skin cancer. This has been studied in goats, cows, cats, dogs, sheep and laboratory animals and is probably pointing out that this is a common feature of several species. Infections in cattle can be aggravated by an increase in UV-B radiation.

Plants

In many plants UV-B radiation can have the following adverse effects such as alter its shape and damage plant growth, reduce tree growth, change flowering times and make plants more vulnerable to disease and produce toxic substances. There could even be losses of biodiversity and species.

Effect on abiotic substances

Destroy the things made up of plastic, rubber, etc.

Some Solutions

- ☆ Replace halon-based fire extinguishers with others using foam.
- ☆ Check on the label of the products, which we buy at the supermarket, to report that they do not damage the ozone layer.
- ☆ Use your car only when necessary. The less we use our cars, the less pollutants we will emit into the atmosphere. Remember that burning of fossil fuels releases many substances that damage the ozone layer.
- ☆ Do not buy refrigerators or air conditioner equipment that uses CFCs as refrigerant. Look for this information in the labels, or ask the supplier of the product directly.
- ☆ Do not use cleaning solvents containing CFCs or ammonia.
- ☆ Do not use sprays, and do not buy objects made of plastic foam (dry ice or freezer). If you receive these products as a fill of your mail packages, return

them immediately to the sender. Low consumption of these products will discourage plastic foam manufacturers.

Montreal Protocol

It is an international and intergovernmental treaty that aimed at reducing the production of CFCs and other ozone depleting substances worldwide. The protocol was signed to protect the depletion of ozone layer. It was registered in September 16, 1987 and was signed on January 1, 1989. It was signed by 196 countries. It aimed to stop the use of halons and CFCs till 2000 and carbon tetra-chloride and chloroform till 2005.

Vienna Convention

On March 22, 1989 Vienna, Austria, a convention for the protection of ozone layer was opened for signature. The convention came into force on 22 September 1989. It was signed by 178 countries till June 2001.

Questions

Short Answer Type Questions

1. What is ozonosphere?
2. What is the height of ozonosphere from Earth's surface?
3. Name the scientists who observed ozone hole for the first time.
4. Where the ozone hole was first reported?
5. Write down the units used to measure ozone thickness.
6. What is the temperature of polar stratospheric clouds?

Essay Type Questions

1. What is ozone layer and write down its importance?
2. What is ozone depletion and how does it occur?
3. What is the connection between ozone depletion and climate change?
4. What is being done to protect ozone layer?
5. Describe various National and International conventions related to OLD

2

Greenhouse Effect

The *greenhouse effect* (GHE) is the process by which radiation from a planet's atmosphere warms the planet's surface to a temperature above what it would be without its atmosphere. Greenhouse effect is what makes this planet suitable for life. Earth's atmosphere contains gases which absorb heat known as greenhouse gases. This heat trapping ability of these greenhouse gases is the reason of greenhouse effect.

According to Goldilocks principle "Venus is too hot (450°C), Mars is too cold (–53°C) and Earth is just right (13°C)". It is not because that our planet orbits at just right distance from sun, but because of the perfect atmosphere. Atmosphere traps heat to keep the global temperature in pleasant range. If the Earth had no atmosphere, its average surface temperature would be very low of about –18°C rather than the comfortable 15°C found today. Due to greenhouse gases, the atmosphere absorbs more infrared energy than it re-radiates to space, resulting in a net warming of the Earth atmosphere system and of surface temperature. This is the "Natural Greenhouse Effect". With more greenhouse gases released to the atmosphere due to human activity, more infrared radiation will be trapped in Earth's surface which contributes to the "Enhanced Greenhouse Effect".

This term was first used in early 1800. At that time it was described as naturally occurring function with not any negative impact. In 1950s, it was coupled with climate change and now it is considered negative due to enhanced greenhouse effect. To understand greenhouse effect, basic facts about solar radiation and structure of greenhouse gases must be known.

Solar Radiation

Sun radiates vast quantities of energy into space, across a wide spectrum of wavelength. Most of the energy is concentrated in visible region (43%). Wavelength shorter than visible account for 7-8%. Remaining 49-50% spread over longer wavelengths. Various components of earth's atmosphere absorb ultra-violet and infrared radiations before it reach the surface but is transparent to visible light.

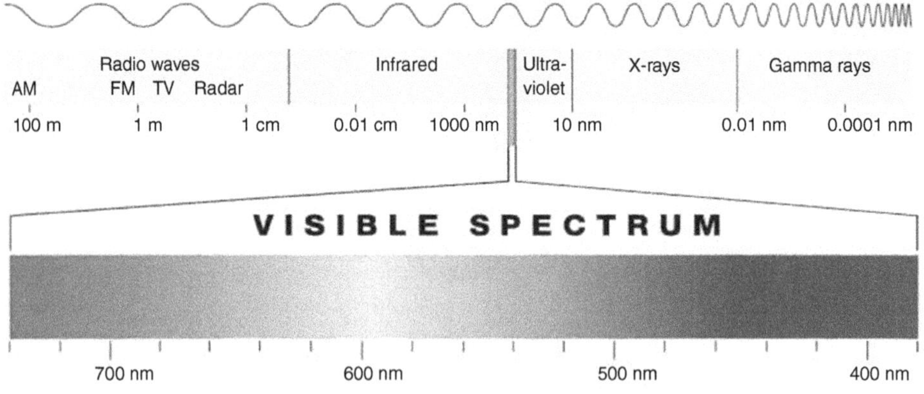

Figure 2.1 Solar spectrum

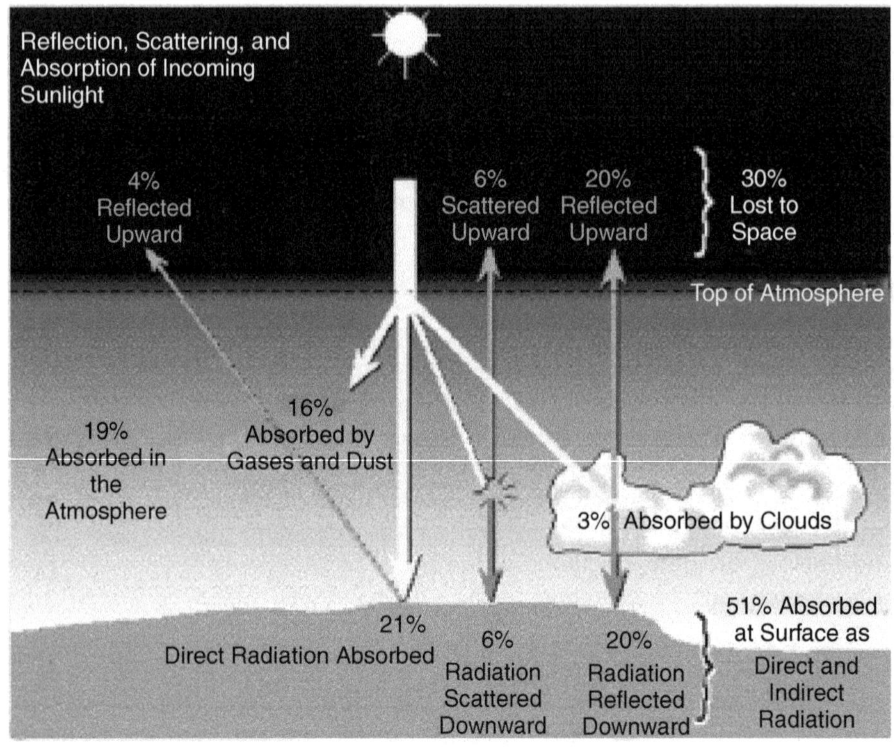

Figure 2.2. Energy budget of Earth

After absorption, visible light is transformed into heat and re-radiates in the form of infrared radiations. If this happens then, in day earth would be hot and at night temperature woud have dropped. This doesn't happen because of the earth's atmosphere contain molecules that absorb heat and known as greenhouse gases.

Greenhouse Gases

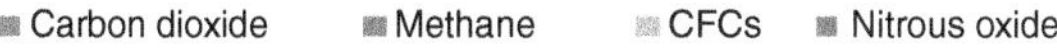

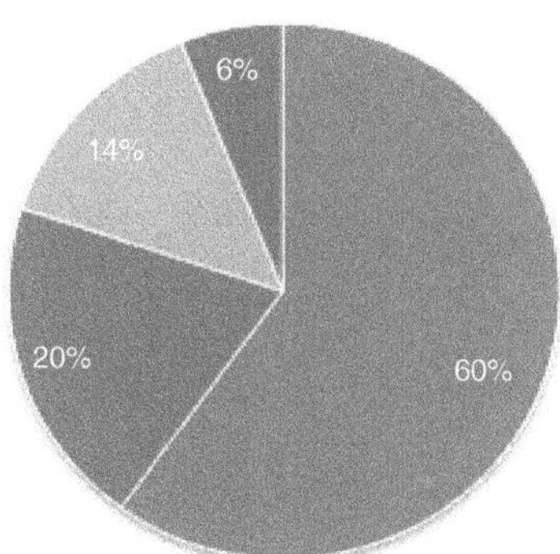

Figure 2.3. The relative contribution of different gases to GHE

There are several different types of greenhouse gases. The major ones are carbon dioxide, water vapour, methane and nitrous oxide. These gas molecules are made up of three or more atoms. The atoms are held together loosely enough that they vibrate when they absorb heat. Vibrating molecule emit radiation again and is absorbed by another greenhouse gas molecule. This absorption-emission-absorption cycle serves to keep heat near the surface. Due to the structure of greenhouse gases this cycle is carried out by them. The major components of atmosphere; oxygen and nitrogen, are too tightly bounded that they can't vibrate, thus do not absorb heat and contribute to greenhouse effect. The major greenhouse gases are discussed below:

Carbon dioxide (CO_2): It contributes about 60% to the global warming. Carbon dioxide enters the atmosphere through burning fossil fuels (coal, natural gas, and oil), solid waste, trees and wood products, and also as a result of certain chemical reactions (e.g., manufacture of cement). Carbon dioxide is removed from the atmosphere (or "sequestered") when it is absorbed by plants as part of the biological carbon cycle.

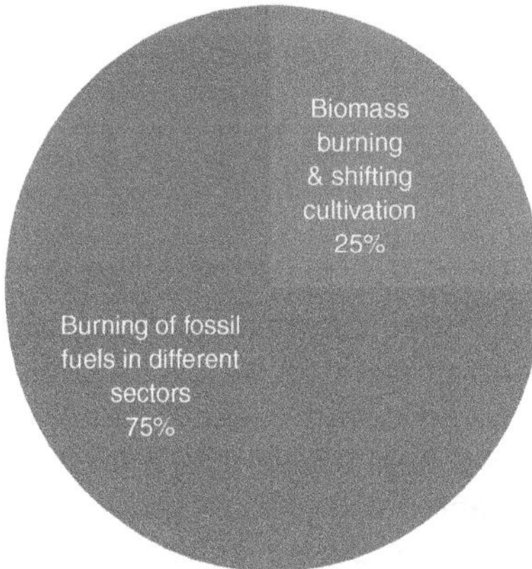

Figure 2.4. Sources of carbon dioxide production

Methane (CH_4): It contributes 20% to the global warming. Methane is emitted during the production and transport of coal, natural gas, and oil. Methane emissions also result from livestock and other agricultural practices and by the decay of organic waste in municipal solid waste landfills.

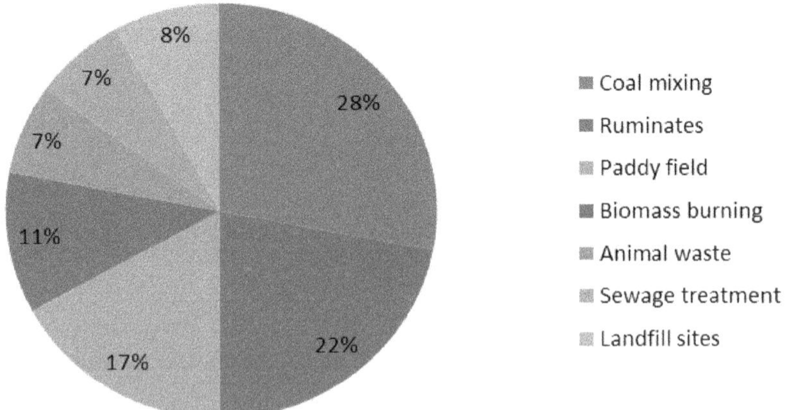

Figure 2.5. Sources of Methane production

Nitrous oxide (N_2O): It is responsible for 6% of global warming. Nitrous oxide is emitted during agricultural and industrial activities, as well as during combustion of fossil fuels and solid waste.

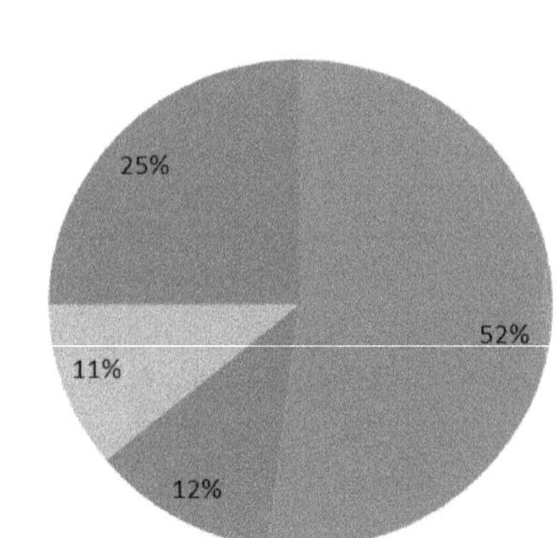

Figure 2.6. Sources of Nitrous Oxide production

Fluorinated gases: Hydrofluorocarbons, perfluorocarbons, sulfur hexafluoride, and nitrogen trifluoride are synthetic, powerful greenhouse gases that are emitted from a variety of industrial processes. Fluorinated gases are sometimes used as substitutes for stratospheric ozone-depleting substances (e.g., chlorofluorocarbons, hydrochlorofluorocarbons, and halons). These gases are typically emitted in smaller quantities, but because they are potent greenhouse gases, they are sometimes referred to as High Global Warming Potential gases ("High GWP gases"). These contribute 14% to global warming.

Causes of Greenhouse Effect

1. **Burning of fossil fuels:** Fossil fuels like coal and natural gas, when burned produce greenhouse gases. Fossil fuels are used to produce electricity and transportation. When these are burnt completely, carbon released combines with oxygen molecule to form carbon dioxide. Due to increase in population vehicles are also increasing, which is increasing carbon dioxide concentration in atmosphere.
2. **Deforestation:** Deforestation, clearance, or clearing is the removal of a forest or stand of trees where the land is thereafter converted to a non-forest use. Plants intake CO_2 and release O_2 through photosynthesis. Large-scale development leads to cutting of trees, due to which there is an increased concentration of CO_2 in the atmosphere, which ultimately leads to enhanced greenhouse effect.
3. **Increase in population:** Increasing population means increased demand of food, cloth and shelter. For fulfilling all these needs of growing population, new manufacturing units are being established which release harmful gases. Moreover, the use of fossil fuels has also increased due to increase in population, which has aggravated the problem of GHE.
4. **Farming:** Nitrous oxide used as fertilizer in the fields is a greenhouse gas and contribute to greenhouse effect. Moreover, rice fields and agricultural soils also release various greenhouse gases which lead to increased greenhouse effect. Field burning of agricultural waste also contributes to greenhouse effect.
5. **Industrial activities and landfills:** Cement industry, chemical industry, fertilizer industry, coal mining, oil extraction produce greenhouse gases. Landfills filled with garbage and solid waste disposal on land produce carbon dioxide and methane.

Consequences of greenhouse effect

The main effect of increased greenhouse gas emissions is global warming. Carbon dioxide, methane, nitrous oxide and fluorinated gases all help to trap heat in the Earth's atmosphere as a part of the greenhouse effect. The Earth's natural greenhouse effect makes life possible on earth. However, human activities, primarily the burning of fossil fuels and deforestation, have intensified the greenhouse effect, causing global warming. Increases in the different greenhouse gases have other effects apart from global warming including ocean acidification, smog pollution, ozone depletion as well as changes to plant growth and nutrition levels.

1. Global Warming

Greenhouse gas levels have been increasing since the start of the Industrial Revolution, but over the last few decades growth has been particularly fast. Total greenhouse gas emissions have increased by about 80% since 1970, creating a radioactive forcing of 2838 mW/m^2 equivalent to an atmospheric concentration of 473 ppm CO_2.

With increasing levels of greenhouse gases being added daily, the greenhouse effect is now enhanced to the point where too much heat is being kept in the Earth's atmosphere. The heat trapped by carbon dioxide and other greenhouse gases has increased surface temperatures by 0.75°C (1.4°F) over the last 100 years. Global warming is harming the environment in several ways including:

- Desertification
- Increased melting of snow and ice
- Sea level rise
- Stronger storms and extreme events

2. Ocean Acidification

Increases in carbon dioxide levels have made the world's oceans 30% more acidic since the Industrial Revolution. The ocean serves as a sink for this gas and absorbs about a quarter of human carbon dioxide emissions, which then goes on to react with sea water to form carbonic acid. So as the level of carbon dioxide in the atmosphere rises, the acidification of the oceans increases.

3. Changes to Plant Growth and Nutrition Levels

Since plants need carbon dioxide to grow, if there are higher amounts in the air, plant growth can increase. Experiments where carbon dioxide concentrations were raised by around 50% increased crop growth by around 15% was observed. Higher levels of carbon dioxide makes carbon more available, but plants also need other nutrients (like nitrogen, phosphorus, etc.) to grow and survive. Without increases in those nutrients as well, the nutritional quality of many plants will decrease. In different experiments with elevated carbon dioxide levels, protein concentrations in wheat, rice, barley, and potato tubers, decreased by 5-14%.

4. Smog and ozone pollution

Over the last century, global background ozone concentrations have become 2 times larger mainly due to increases in methane and nitrogen oxides caused by human emissions. At ground level, ozone is an air pollutant that is a major component of smog which is dangerous for both humans and plants. Long-term ozone exposure has also been shown to reduce life expectancy. 362000-700000 of annual premature cardiopulmonary deaths worldwide are attributable to ozone. Recent studies estimate that the global yields of key staple crops, like soybean, maize (corn), and wheat, are being reduced by 2-15% due to present-day ozone exposure.

5. Ozone layer depletion

Nitrous oxide damages the ozone layer and is now the most important ozone depleting substance and the largest cause of ozone layer depletion. This is because CFCs and many other gases that are harmful for the ozone layer were banned by the Montreal Protocol (MP) which has reduced their atmospheric concentration. Nitrous oxide is not restricted by the MP, so while the levels of other ozone depleting substances are declining, nitrous oxide levels are continuing to grow.

Questions

Short Answer Type Questions

1. Define enhanced greenhouse effect.
2. What is the average surface temperature of Earth?
3. Name the portion of solar spectrum to which earth's atmosphere is transparent.
4. What are the natural causes of GHE?

Essay Type Questions

1. What is greenhouse Effect? How does it work?
2. What are greenhouse gases and list natural and anthropogenic sources of different greenhouse gases?
3. Explain the causes of greenhouse effect.
4. Describe in details the consequences of GHE.

3

Population Growth

Population is a group of organisms of a particular species occupying a particular area at a specific time. In biology or human geography, ***population growth*** is the increase in the number of individuals in a population.

Global human population growth amounts to around 83 million annually, or 1.1% per year. The global population has grown from 1 billion in 1800 to 7.616 billion in 2018. It is expected to keep growing, and estimates have put the total population at 8.6 billion by mid-2030, 9.8 billion by mid-2050 and 11.2 billion by 2100. Many nations with rapid population growth have low standards of living, whereas many nations with low rates of population growth have high standards of living.

The rapid growth of the world's population over the past one hundred years results from a difference between the rate of birth and the rate of death. The human population will increase by 1 billion people in the next decade. This is like adding the whole population of China to the world's population. The growth in human population around the world affects all people through its impact on the economy and environment. The current rate of population growth is now a significant burden to human well-being. Understanding the factors which affect population growth patterns can help us plan for the future.

Population growth is not a problem until resources are available. Presently population is increasing and that is root of many problems. Increase in population leads to increased demand for food, shelter, resources, etc. More waste is generated and pollution is also elevated due to increased population. This is further worsened with the increased competition for resources. Therefore population should be controlled.

Characteristics of Population

- ☆ **Birth rate:** Number of births in particular population in particular time is known as birth rate.

- **Death rate:** Number of deaths at particular time in particular population is known as death rate.
- **Age distribution:** In every population there are different members with different age and members with same age are collectively known as age group. Population increase in any population depends on the distribution of different age groups in the population. Basically, in every population there are three types of age groups:
- Pre-reproductive
- Reproductive
- Post-reproductive

To understand the age distribution, different age pyramids are employed. Population pyramids are graphical representations of the age of a population. There are three types of population pyramids:

1. Expanding Pyramid

It depicts the populations that have larger percentages of people in younger age groups. Populations with this shape usually have high fertility rates. In India U.P. is having expanding type population pyramid.

2. Stationery Pyramid

These pyramids show somewhat equal proportion of the population in each group. There is not a decrease or increase in the population; it is stable. Kerela is having stationary type of population pyramid.

3. Contracting Pyramid

There is a lower percentage of younger people. It shows declining birth rates since each succeeding age group is getting smaller and smaller.

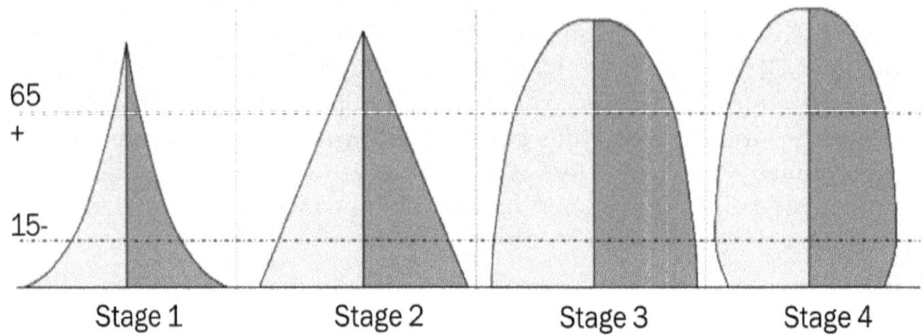

Expanding pyramid **Stationery Pyramid** **Contracting Pyramid**

Figure 3.1. Different types of Population pyramids

In 14th century, 50% population in Europe and Asia was destroyed due to plague; and there was a great decrease in the human population. But, latter on, in around 18th century due to scientific development and increased health facility population was increased. According to population increase, there are four stages of population.

1. Stable Stage

This was the first stage of population and it ranged from ancient time to 18th century. At that time life was difficult due to limited health facility. Agriculture activities were initiated in this stage which leads to increased birth rate, but death rate and baby death rate were high. So population increase was limited.

2. Transitional Stage

This stage ranged beyond 18th century and during this stage, due to industrialization and scientific development death rate was decreased. Since there was no lack of food and other resources, control over increasing population was not done. As a result death rate was decreased, therefore population was increased rapidly.

3. Industrial Stage

In this stage, industrialization and scientific development was increased due to which death rate was decreased. Further resources were affected and population pressure was observed. Due to the visible effects of increased population, birth rate was controlled and was decreased but not more than death rate. Therefore, population increase was observed in this stage but the increase was slow.

4. Industrialized Stage

This is the last or extreme stage of population growth. Here the birth rate and death rate were controlled due to improved health facilities. Therefore no increase in population was observed and known as zero population growth.

In India, due to scientific development, death rate was decreased, but birth rate is not decreased. Therefore, population is increasing and it falls in industrial stage. It is the 2nd largest population of world and it will take 50 years to achieve zero population stage.

Overpopulation

Overpopulation is an undesirable condition where the number of existing human population exceeds the carrying capacity of Earth. Overpopulation is caused by number of factors. Reduced mortality rate, medical facilities, depletion of precious resources are few of the causes which result in overpopulation. It is possible for a sparsely populated area to become densely populated if it is not able to sustain life.

Growing advances in technology with each coming year has affected humanity in many ways. One of these has been the ability to save and create better medical treatment for all. A direct result of this has been increased lifespan and the growth of the population. In the past fifty or so years, the growth of population has boomed and has turned into overpopulation. Developing nations face the problem of overpopulation more than developed countries, but it affects most of the Earth as of now.

Causes of Overpopulation

1. Decline in the death rate

At the root of overpopulation is the difference between the overall birth rate and death rate in populations. If the number of children born each year equals the number of adults that die, then the population will stabilize. There are many factors that can increase the death rate for short periods of time, the ones that increase the birth rate do so over long period of time. The discovery of agriculture was one factor that provided them with the ability to sustain their nutrition without hunting which created the first imbalance between the two rates.

2. Better Medical Facilities

Following this came the industrial revolution. Technological advancement was perhaps the biggest reason why the balance has been permanently disturbed. Science was able to produce better means of producing food, which allowed families to feed more mouths. Medical science made many discoveries due to which a whole range of diseases were defeated. Combining the increase in food supply with fewer means of mortality tipped the balance and became the starting of overpopulation.

3. More Hands to Overcome Poverty

However, when talking about overpopulation we should understand that there is a psychological components as well. For thousands of years, a very small part of the population had enough money to live in comfort. The rest faced poverty and gave birth to large families to make up for the high infant mortality rate. Families that have been through poverty, natural disaster or are simply in need of more hands to work are a major factor for overpopulation.

4. Technological Advancement in Fertility Treatment

With the latest technological advancement and more discoveries in medical science, it has become possible for couple who are unable to conceive, to ungergo fertility treatment methods and have their own babies. Today there are effective medicines which can increase the chance of conception and lead to rise in birth rate. Moreover, due to modern techniques pregnancies today are far more safer.

5. Immigration

Many people prefer to move to developed countries like US, UK, Canada and Australia where best facilities are available in terms of medical, security and employment. The end result is that people settle over there and those places become overcrowded. Difference between the number of people who are leaving the country and the number of people who entered narrow towns which leads to more demand for food, clothes, energy and homes. This gives rise to shortage of resources.

3. Lack of Family Planning

Most of developing nations have large number of people who are illiterate, live below the poverty line and have little or no knowledge about family planning. Getting their children married at an early age increase the chances of produing more kids.

Effects of Overpopulation

1. Economic Effects

Inflation: As a difference between demand and supply continues to expand due to overpopulation, it raises the prices of various commodities including food, shelter and healthcare. This means that people have to pay more to survive and feed their families.

Increased unemployment: When a country becomes overpopulated, it gives rise to unemployment as there fewer jobs to support large number of people.

Increased poverty: Poverty is the biggest hallmark we see when talking about overpopulation. All of this only becomes worse if solutions are not sought out for the factors affecting our population.

2. Social Effects

Crime: Decrease in natural resources and competition for food and shelter combined with unemployment, creates the feeling of dissatisfaction which leads to increased crime in the society. This ultimately leads to social fragmentation.

Conflicts and wars: Overpopulation in developing countries puts a major strain on the resources it should be utilizing for development. Conflicts over water are becoming a source of tension between countries, which could result in wars.

3. Environmental Effects

Depletion of natural resources: Earth can only produce a limited amount of water, food, which is falling short of the current needs. Cutting down of forests, hunting wildlife in a reckless manner is creating a host of problems.

Increased pollution: Increased population leads to elevated levels of environmental pollution due to the overuse of fossil fuels like coal and petroleum. Rise in number of vehicles and industries have badly affected the quality of air.

Global warming: Rise in amount of CO_2 emissions leads to global warming. Melting of polar ice caps, changing climate patterns, rise in sea level are few of the consequences that we might have to face due to environmental pollution.

Solutions to Overpopulation

1. **Better education**: One of the first measures is to implement policies reflecting social change. Educating the masses helps them understand the need to have one or two children at the most. Similarly, education plays a vital role in understanding latest technologies that are making huge waves in the world of computing. Families that are facing a hard life and choose to have four or five children should be discouraged. Family planning and efficient birth control can help in women making their own reproductive choices. Open dialogue on abortion and voluntary sterilization should be seen when talking about overpopulation.
2. **Making people aware of family planning:** As population of this world is growing at a rapid pace, raising awareness among people regarding family planning and letting them know about serious after effects of overpopulation can help curb population growth. One of the best way is to let them know about various safe sex techniques and contraceptives methods available to avoid any unwanted pregnancy.

3. **Tax benefits or concessions:** Government of various countries might have to come with various policies related to tax exemptions to curb overpopulation. One of them might be to waive off certain part of income tax or lowering rates of income tax for those married couples who have single or two children. As we humans are more inclined towards money, this may produce some positive results.
4. **Knowledge of sex education:** Imparting sex education to young kids at elementary level should be must. Most parents feel shy in discussing such things with their kids which result in their children going out and look out for such information on internet or discuss it with their peers. Mostly, the information is incomplete which results in sexually active teenagers unaware of contraceptives and embarrassed to seek information about same. It is, therefore, important for parents and teachers to shed their old inhibitions and make their kids or students aware of solid sex education.

The rapid growth of world's population over the past one hundred years results from a difference between the rate of birth and the rate of death. The growth in human population around the world affects all people through its impact on the economy and environment.

The current rate of population growth is now a significant burden to human well-being. Understanding the factors which affect population growth patterns can help us plan for the future.

Questions

Short Answer Type Questions

1. Define population.
2. What is actual population growth?
3. What is India's rank in world in terms of human population?

Essay Type Questions

1. What are the characteristics of a population?
2. Explain various stages of population growth.
3. What is population explosion and its causes?
4. Describe the effects of human population growth.

4

Industrialisation

Industrialisation or ***industrialization*** is the period of social and economic change that transforms a human group from an agrarian society into an industrial society, involving the extensive re-organization of an economy for the purpose of manufacturing. It can be defined as development of industries in a country or in a region on a wide scale.

As industrial workers' incomes rise, markets for consumer goods and services of all kinds tend to expand and provide a further stimulus to industrial investment and economic growth.

The first transformation to an industrial economy from an agricultural one, known as the Industrial Revolution, took place from the mid-18th to early 19th century. By the end of the 20th century, East Asia had become one of the most recently industrialized regions of the world.

Causes

1. Abundance of Natural Resources

Abundance of natural resources in any region is also a big reason behind the spread of industrialization. More is the availability of natural resources in any region, cheaper is the raw material for the industries. Like, forest resources invite a number of industries for obtaining raw material in industries like paper and pulp, timber industry, etc. Therefore, when industries are installed in forest the waste is also dumped into the nature. For example waste water generated from paper industry contain sodium hyroxide, Sodium carbonate, Chlorine dioxide, elemental chlorine. Similarly in cement industry cement dust or flyash is released into the environment which contain heavy metals like Ni, Co, Pb and Cr.

2. Urbanization

The concentration of labour into factories has increased urbanization and the size of settlements, to serve and house the factory workers. Moreover, the improved facilities such as better education, health care, sanitation, housing, business and transportation also calls a huge population from rural areas to urban areas.

3. Growing Population

Due to the increase in population, the demand and need of goods and other material is also increased. This leads to the growth of industries in that particular area so that the demand of the increasing population could be fulfilled.

4. Exploitation

Workers have to either leave their families or bring them along in order to work in the towns and cities where these industries are found. Due to this, a large population is forced to migrate to the cities.

5. Changes in Family Structure

The family structure changes with industrialization. In industrialized societies the nuclear family, consisting of only parents and their growing children, predominates. Families and children reaching adulthood are more mobile and tend to relocate to where jobs exist. Extended family bonds become more tenuous.

Current Situation

The relationships among economic growth, employment, and poverty reduction are complex. Higher productivity, it is argued, may lead to lower employment (see jobless recovery). There are differences across sectors, whereby manufacturing is less able than the tertiary sector to accommodate both increased productivity and employment opportunities; more than 40% of the world's employees are "working poor", whose incomes fail to keep themselves and their families above the $2-a-day poverty line. There is also a phenomenon of de-industrialization.

Air pollution, greenhouse gases emission, global warming, climatic disasters, water shortages, drinking water contamination, freshwater and marine pollution, deforestation, and other environmental problems are becoming serious threats to the well-being of mankind in this age of industrialization. There are various wide-ranging effects, as well as serious consequences, of industrial pollution on the ecological balance of the atmosphere.

Global warming is one of the most common and serious consequence of industrial pollution, causing an increase of the water levels in seas and rivers, thereby increasing the chances of flood. With the increase in the number of industries and factories due to the industrial revolution; air pollution also has increased significantly. The emissions from various industries contain large amounts of gases, which, when present in elevated levels in the atmosphere, often result in various environmental and health hazards such as acid rain, and various skin disorders in individuals. Dumping of various industrial waste products into water sources, and improper treatment of industrial wastes, often result in polluting the water disturbing the balance of the ecosystem inside, resulting in the death of various animal and plant species present in the water.

Certain other common effects of industrial pollution include damaging buildings and structures, increasing the risk of various occupational hazards such as asbestosis, pneumoconiosis, among others. This chapter provides the true global assessment of the scale of the direct impacts of industrialization on human environment and atmosphere, the steps that can be taken to curb industrial pollution and policy recommendations for sustainability.

India is a developing nation. India is well thought-out as the world's biggest booming economies. Modernization has led to the development in the lifestyle and the basic needs are no more just food, cloth and shelter. The industrialization has led to development in diverse areas like agriculture, manufacturing sector, coal, timber, bottling plants, automobiles, gas and chemicals. This has definitely developed the economy of India and the lifestyle of people living in the country. It had also led to the degradation of environment and the environmental conditions, the flora and fauna in different ecosystems, extinction of the rare species of animals, plants and birds and the depletion of natural resources.

The major cause of this is the deforestation of the ecosystems for industrialization. Industrialization has resulted in the increase in the emission of harmful effluents and pollutants both into water, soil and air. These effluents have caused a severe and irreversible destruction to the different species residing in those specific ecosystems. Many life species have become vulnerable and some are extinct due to deforestation. The groundwater reserves are no more pure, global warming has caused in the depletion of the ozone layer and has caused deadly diseases in certain areas. Most of this is caused by heavy industrialization. The ill effects of industrialization are very well known but the major question remains whether we want this to continue and suffer in the near future or make a change.

Industrial activities are one of the major sources of air, water and land pollution in India. World Health Organization estimates that outdoor air pollution only accounts for around 2% of all heart and lung diseases, 5% of all lung cancers, and about 1% of all chest infections globally. The recent changes in Indian political system suits industrial growth, but at the similar time we need to be more careful towards industrial hazards. Even after witnessing one of the most horrible industrial disasters of all time in Bhopal in 1984, there has not been much improvement in industrial sector. In the three days, around 8,000 people died.

Thousands of people still feel the effects even after three decades. It is evident that pollution from industries has pessimistic impact resulting in loss of unique genetic resources. This is a evolution period for many developing economies like India, so there is a strong call to strike a balance between industrial growth and physical environment so as to reduce the intensity of pollution. In this paper an attempt has been made to analyze the ill effects of industrialization on the environmental pollution.

Effects

Positive Effects

- ☆ Low cost of production
- ☆ Less time
- ☆ Self-dependent
- ☆ Employment
- ☆ Improved agriculture

Side Effects

- **Depletion of natural resources:** Industrialization leads to depletion of natural resources in the area of industrial establishment. Metal contaminants like Cd, Zn, Hg, etc., destroy bacteria and beneficial micro-organisms in the soil.
- **Pollution:** It leads to air pollution, water pollution and soil pollution.
- **Global warming and climatic changes:** Global warming and climatic changes are the major consequences of industrialization.
- **Acid rain:** More gaseous air pollutants are released in the atmosphere, which further rise in the atmosphere and falls back to earth surface in the form of acid rain.
- **Degradation of land quality:** The solid waste discharge of the industries results in degradation of land quality where it is disposed. Industrial effluents damages several soil and waterborne diseases.
- **Waste generation:** It leads to generation of hazardous waste whose safe disposal become a big problem. Industrial wastes including toxins entering the food chain causes number of undesirable effects to living beings and animals which include silicosis and pneumoconiosis, tuberculosis, skin diseases and deafness. Radioactive industrial pollutant cause undesirable diseases when food containing radio-nuclides is taken by man.

Questions

Short Answer Type Questions

1. What are major pollutants from cement industry?
2. What is industrial revolution?
3. What is industrialization?
4. Name the major pollutants from sugar and paper industry.

Essay Type Questions

1. What are the factors contributing to industrialization?
2. In what ways industrialization is changing our society?
3. What are the long-term consequences of industrialization?
4. Name some major industries of India and the waste generated by them.

5

Urbanization

Urbanization is a process whereby populations move from rural to urban area, enabling cities and towns to grow. It can also be termed as the progressive increase of the number of people living in towns and cities. In other words, we can say that it is a process that leads to the growth of cities due to industrialization and economic development. It is highly influenced by the notion that cities and towns have achieved better economic, political, and social mileages compared to the rural areas.

Accordingly, urbanization is very common in developing and developed worlds as more and more people have the tendency of moving closer to towns and cities to acquire privileged social and economic services as well as benefits. These include social and economic advantages such as better education, health care, sanitation, housing, business opportunities and transportation.

Majority of people move to cities and towns because they view rural areas as places with hardship and backward or primitive lifestyle. Therefore, as populations move to more developed areas the immediate outcome is urbanization. This normally contributes to the development of land for use in commercial properties, social and economic support institutions, transportation, and residential buildings. Eventually, these activities raise several urbanization issues.

Causes of Urbanization

1. Industrialization

Industrialization is a trend representing a shift from the old agricultural economics to novel non-agricultural economy, which creates a modernized society. Through industrial revolution, more people have been attracted to move from rural to urban areas on the account of improved employment opportunities. Industrialization has increased employment opportunities by giving people the chance to work in modern sectors in job categories that aids to stir economic developments.

2. Commercialization

Commerce and trade play a major role in urbanization. The distribution of goods and services and commercial transactions in the modern era has developed modern marketing institutions and exchange methods that have tremendously given rise to the growth of towns and cities. Commercialization and trade comes with the general perception that the towns and cities offer better commercial opportunities and return compared to the rural areas.

3. Social Benefits and Services

There are numerous social benefits attributed to life in the cities and towns. Examples include better educational facilities, better living standards, and better sanitation and housing, better health care, better recreation facilities and better social life in general. On this account, more and more people are prompted to migrate into cities and towns to obtain the wide variety of social benefits and services which are unavailable in rural areas.

4. Employment Opportunities

In cities and towns, there are ample job opportunities that continually draw people from the rural areas to seek better livelihood. Therefore, the majority of people frequently migrate into urban areas to access well paying jobs as urban areas have countless employment opportunities in all developmental sectors such as public health, education, transport, sports and recreation, industries and business enterprises. Services and industries generate and increase higher value-added jobs and this leads to more employment opportunities.

5. Modernization and Changes in the Mode of Living

Modernization plays a very important role in the process of urbanization. As urban areas become more technology savvy together with highly sophisticated communication, infrastructure, medical facilities, dressing code, enlightenment, liberalization and other social amenities availability, people believe they can lead a happy life in cities. In urban areas, people also embrace changes in the modes of living namely residential habits, attitudes, dressing, food and beliefs. As a result, people migrate to cities and the cities grow by absorbing the growing number of people day after day.

6. Rural Urban Transformation

As localities become more fruitful and prosperous due to the discovery of minerals, resource exploitation, or agricultural activities, cities start emerging as the rural areas transform to urbanism. The increase in productivity leads to economic growth and higher value-added employment opportunities.

This brings about the need to develop better infrastructure, better education institutions, better health facilities, better transportation networks, establishment of banking institutions, better governance and better housing. As this takes place, rural communities start to adopt the urban culture and ultimately become urban centers that continue to grow as more people move to such locations in search of a better life.

Effects of Urbanization

I. Positive Effects of Urbanization

Urbanization yields several positive effects if it happens within the appropriate limits. Some of the positive implications of urbanization, therefore, include creation of employment opportunities, technological and infrastructural advancements, improved transportation and communication, quality educational and medical facilities and improved standards of living. However, extensive urbanization mostly results in adverse effects. Few negative effects are listed below.

II. Negative Effects of Urbanization

1. Housing Problems

Urbanization attracts people to cities and towns which lead to high population increase. With the increase in the number of people living in urban centers, there is continued scarcity of houses. This is due to insufficient expansion space housing and public utilities, poverty, unemployment and costly building materials which can only be afforded by few individuals.

2. Overcrowding

Overcrowding is a situation whereby a huge number of people live in a small space. This form congestion in urban areas is consistent because of overpopulation and it is an aspect that increases day by day as more people and immigrants move into cities and towns in search of better life. Most people from rural or undeveloped areas always have the urge of migrating into the city that normally leads to congestion of people within a small area.

3. Unemployment

The problem of joblessness is highest in urban areas and it is even higher among the educated people. It is estimated that more than half of unemployed youths around the globe live in metropolitan cities. And, as much as income in urban areas is high, the costs of living make the incomes to seem horribly low. The increasing relocation of people from rural or developing areas to urban areas is the leading cause of urban unemployment.

4. Development of Slums

The cost of living in urban area is very high. When this is combined with random and unexpected growth as well as unemployment, there is the spread of unlawful resident settlements represented by slums and squatters. The growth of slums and squatters in urban areas is even further exacerbated by fast-paced industrialization, lack of developed land for housing, large influx of rural immigrants to the cities in search of better life and the elevated prices of land beyond the reach of the urban poor.

5. Water and Sanitation Problems

Because of overpopulation and rapid population increase in most urban centres, it is common to find there are inadequate sewage facilities. Municipalities and local governments are faced with serious resource crisis in the management of sewage facilities. As a result, sanitation becomes poor and sewages flow chaotically and they are drained into neighbouring streams, rivers, lakes or seas. Eventually, communicable diseases such as typhoid, dysentery, plague and diarrhea spread very fast leading to suffering and even deaths. Overcrowding also highly contributes to water scarcity as supply falls short of demand.

6. Poor Health and Spread of Diseases

The social, economic and living conditions in congested urban areas affects access and utilization of public health care services. Slum areas in particular experience poor sanitation and insufficient water supply which generally make slum populations susceptible to communicable diseases. The environment problems such as urban pollution also cause many health problems namely allergies, asthma, infertility, food poisoning, cancer and even premature deaths.

7. Traffic Congestion

When more people move to towns and cities, one of the major challenges posed is in the transport system. More people means increased number of vehicles which lead to traffic congestion and vehicular pollution. Many people in urban areas drive to work and this creates a severe traffic problem, especially during the rush hours. Also as the cities grow in dimension, people will move to shop and access other social needs which often cause traffic congestion and blockage.

8. Urban crime

Issues of lack of resources, overcrowding, unemployment, poverty and lack of social services and education habitually leads to many social problems including violence, drug abuse and crime. Most of the crimes such as murder, rape, kidnapping, riots, assaults, theft, robbery and hijacking are reported to be more prominent in the urban vicinities. Besides, poverty related crimes are the highest in fast-growing urban regions. These acts of urban crime normally upset the peace and tranquility of cities and towns.

Solutions of Urbanization

1. Building Sustainable and Environmentally Friendly Cities

Governments should pass laws that plan and provide environmentally sound cities and smart growth techniques, considering that people should not reside in unsafe and polluted areas. The objective here is to build sustainable cities that embrace improved environmental conditions and safe habitats for all urban populations. Government should also encourage sustainable use of urban resources and support an economy based on sustainable environment such as investment in green infrastructure, sustainable industries, recycling and environmental campaigns, pollution management, renewable energy, green public transportation and water recycling and reclamation.

2. Provision of essential services

Urban stakeholders must ensure all populations within the urban areas have access to adequate essential social services namely education, health, sanitation and clean water, technology, electricity and food. The objective here is to provide and implement employment opportunities and wealth creation activities so that people can earn a living to pay for the maintenance of the services. Subsidies can also be availed by the government to lower the costs of basics health care, basic education, energy, education, public transportation, communication system and technology.

3. Creation of More Jobs

To lessen the negative effects of rapid urbanization while at the same time conserving natural ecosystems, private investments should be encouraged so as to utilize natural resources and create more job opportunities. Tourism promotion and the sustainable exploitation of natural resources can create more jobs for the urban populations. Subsidies and grants may be provided to foreign and private investment in environmentally friendly development projects that encourage job creation.

4. Population Control

Key stakeholders in urban areas must provide campaigns and counseling for effective medical health clinics and family planning to help reduce the high rates of population growth. Medical health clinics oriented towards family planning options must be made accessible across the entire urban area with the objective of controlling diseases and population growth.

Questions

Short Answer Type Questions

1. Dfine urbanization.
2. What is commercialization?
3. What is initial urban transformation?

Essay Type Questions

1. What are the most prominent causes of urbanization?
2. How does resource extraction effect urbanization?
3. How much urbanization as a factor is responsible for climate change?
4. What are the impacts of urbanization?
5. What are the efficient methods which can be adopted so that side effects of urbanization can be minimized?

6

Environmental Impact of Modern Agriculture

The environmental impact of agriculture is the effect that different farming practices have on the ecosystems around them, and how those effects can be traced back to those practices. The environmental impact of agriculture varies based on the wide variety of agricultural practices employed around the world. Ultimately, the environmental impact depends on the production practices of the system used by farmers. The connection between emissions into the environment and the farming system is indirect, as it also depends on other climate variables such as rainfall and temperature.

Modern Agriculture

Modern agriculture is the term used to describe the wide type of production practices such as the use of hybrid seeds of selected varieties of a single crop, technologically advanced equipment and lots of energy subsides in the form of irrigation water, fertilizers and pesticides. Modern agriculture includes:

1. High yielding varieties
2. Improved irrigation practices
3. New fertilizers
4. Insect resistant varieties
5. Diseases resistant varieties
6. New instruments and machines

The environmental impact of agriculture involves a variety of factors from the soil, to water, the air, animal and soil variety, people, plants, and the food itself. Some of the environmental issues that are related to agriculture are climate change, deforestation, genetic engineering, irrigation problems, pollutants, soil degradation, and waste.

Impacts of Modern Agriculture

1. Climate Change

Climate change and agriculture are interrelated processes, both of which take place on a worldwide scale. Global warming is projected to have significant impacts on conditions affecting agriculture, including temperature, precipitation and glacial run-off. These conditions determine the carrying capacity of the biosphere to produce enough food for the human population and domesticated animals. Rising carbon dioxide levels would also have effects, both detrimental and beneficial, on crop yields. Assessment of the effects of global climate changes on agriculture might help to properly anticipate and adapt farming to maximize agricultural production. Although the net impact of climate change on agricultural production is uncertain, it is likely that it will shift the suitable growing zones for individual crops. Adjustment to this geographical shift will involve considerable economic costs and social impacts.

At the same time, agriculture has been shown to produce significant effects on climate change, primarily through the production and release of greenhouse gases such as carbon dioxide, methane, and nitrous oxide. In addition, agriculture that practices tillage, fertilization, and pesticide application also releases ammonia, nitrate, phosphorus, and many other pesticides that affect air, water, and soil quality, as well as biodiversity. Agriculture also alters the Earth's land cover, which can change its ability to absorb or reflect heat and light, thus contributing to radioactive forcing. Land use change such as deforestation and desertification, together with use of fossil fuels, are the major anthropogenic sources of carbon dioxide; agriculture itself is the major contributor to increasing methane and nitrous oxide concentrations in earth's atmosphere.

2. Hazard to Biodiversity

Due to deforestation, initially, total number of species present in an area is decreased. Secondly, due to monoculture similar crops are grown which further result in loss of biodiversity.

3. Depletion of Ecosystem

For agriculture land is needed, and that is obtained by clearing the forests. Deforestation leads to the total destruction of the forest ecosystem. Deforestation is clearing the Earth's forests on a large-scale worldwide and resulting in many land damages. One of the causes of deforestation is to clear land for pasture or crops. According to British environmentalist Norman Myers, 5% of deforestation is due to cattle ranching, 19% due to over-heavy logging, 22% due to the growing sector of palm oil plantations, and 54% due to slash-and-burn farming.

Deforestation causes the loss of habitat for millions of species, and is also a driver of climate change. Trees act as a carbon sink: that is, they absorb carbon dioxide, an unwanted greenhouse gas, out of the atmosphere. Removing trees releases carbon dioxide into the atmosphere and leaves behind fewer trees to absorb the increasing amount of carbon dioxide in the air. In this way, deforestation accelerates climate change. When trees are removed from forests, the soils tend to dry out because there is no longer shade, and there are not enough trees to assist in the water cycle by returning water vapour back to the environment. With no trees, landscapes that were once forests can potentially become barren deserts. The removal of trees also causes extreme fluctuations in temperature.

4. Genetic Engineering

Genetically engineered crops are herbicide-tolerant, and their overuse has created herbicide resistant "super weeds", which may ultimately increase the use of herbicides. To increase the yields, farmers are opting for high yielding varieties. Seed contamination is another problem of genetic engineering; it can occur from wind or bee pollination that is blown from genetically-engineered crops to normal crops. About 50% of corn and soybean samples and more than 80% of canola samples were found to be contaminated by Monsanto's (genetic engineering company) genes. This accidental contamination can cause organic farmers to lose a lot of money because they needed to recall their products. There are various cases of this such as in the corn and alfalfa industry. This leads to monoculture, putting excess pressure on soil and causing deficiency of selective nutrients in soil.

5. Irrigation

Irrigation can lead to a number of problems: Among some of these problems is the depletion of underground aquifers through over drafting. Soil can be over-irrigated because of poor distribution uniformity or management of waste water, chemicals, and this may lead to water pollution. Over-irrigation can cause deep drainage from rising water tables that can lead to problems of irrigation salinity requiring water table control by some form of subsurface land drainage. However, if the soil is under irrigated, it gives poor soil salinity control which leads to increased soil salinity with consequent buildup of toxic salts on soil surface in areas with high evaporation. This requires either leaching to remove these salts and a method of drainage to carry the salts away. Irrigation with saline or high-sodium water may damage soil structure owing to the formation of alkaline soil.

6. Pollutants

Synthetic pesticides such as 'Malathion', 'Rogor', 'Kelthane' and 'confidor' are the most widespread method of controlling pests in agriculture. Pesticides can leach through the soil and enter the groundwater, as well as linger in food products and result in death in humans. Pesticides can also kill non-target plants, birds, fish and other wildlife. A wide range of agricultural chemicals are used and some become pollutants through use, misuse, or ignorance. The erosion of topsoil, which can contain chemicals such as herbicides and pesticides, can be carried away from farms to other places. Pesticides can be found in streams and groundwater.

Atrazine is a herbicide used to control weeds that grow among crops. This herbicide can disrupt endocrine production which can cause reproductive problems in mammals, amphibians and fish that have been exposed. Pollutants from agriculture have a huge effect on water quality. Agricultural nonpoint source (NPS) impacts lakes, rivers, wetlands, estuaries, and groundwater and leads to the problem of eutrophication. Agricultural NPS can be caused by poorly managed animal feeding operations, overgrazing, plowing, fertilizer, and improper, excessive, or badly timed use of pesticides. Pollutants from farming include sediments, nutrients, pathogens, pesticides, metals, and salts. Bacteria and pathogens in manure can make their way into streams and groundwater if grazing, storing manure in lagoons and applying manure to fields is not properly managed.

7. Soil Degradation and Soil Erosion

In modern agriculture, there is an increased use of chemical fertilizers and pesticides so that the crop production could be increased, which leads to decreased soil fertility. This is further accelerated when more than one crop per year, further worsening the status of soil. Ultimately the soil is converted to barren land forever.

Soil degradation is the decline in soil quality that can be a result of many factors, especially from agriculture. Soils hold the majority of the world's biodiversity, and healthy soils are essential for food production and an adequate water supply. Common attributes of soil degradation can be salting, waterlogging, compaction, pesticide contamination, decline in soil structure quality, loss of fertility, changes in soil acidity, alkalinity, salinity, and erosion. Soil erosion is the wearing away of topsoil by water, wind, or farming activities. Topsoil is very fertile, which makes it valuable to farmers growing crops. Soil degradation also has a huge impact on biological degradation, which affects the microbial community of the soil and can alter nutrient cycling, pest and disease control, and chemical transformation properties of the soil.

8. Waste

Plasticulture is the use of plastic mulch in agriculture. Farmers use plastic sheets as mulch to cover 50-70% of the soil and allow them to use drip irrigation systems to have better control over soil nutrients and moisture. Rain is not required in this system, and farms that use plasticulture are built to encourage the fastest runoff of rain. The use of pesticides with plasticulture allows pesticides to be transported easier in the surface runoff towards wetlands or tidal creeks. The runoff from pesticides and chemicals in the plastic can cause serious deformations and death in shellfish as the runoff carries the chemicals towards the oceans.

In addition to the increased runoff that results from plasticulture, there is also the problem of the increased amount of waste form the plastic mulch itself. The use of plastic mulch for vegetables, strawberries, and other orchard crops exceeds 110 million pounds annually in the United States. Most plastic ends up in the landfill, although there are other disposal options such as disking mulches into the soil, on-site burying, on-site storage, reuse, recycling, and incineration. The incineration and recycling options are complicated by the variety of the types of plastics that are used and by the geographic dispersal of the plastics. Plastics also contain stabilizers and dyes as well as heavy metals, which limits the amount of products that can be recycled. Research is continually being conducted on creating biodegradable or photodegradable mulches. While there has been minor success with this, there is also the problem of how long the plastic takes to degrade, as many biodegradable products take a long time to break down.

9. Mechanized Farming

Advances in technologies lead to the use of mechanized farming increases input costs and use of fossil fuel to run the machines also adds to atmospheric pollution.

Sustainable Agriculture

Sustainable agriculture is the idea that agriculture should occur in a way such that we can continue to produce what is necessary without infringing on the ability for future generations to do the same.

The exponential population increase in recent decades has increased the practice of agricultural land conversion to meet demand for food which in turn has increased the effects on the environment. The global population is still increasing and will eventually stabilize, as some critics doubt that food production, due to lower yields from global warming, can support the global population. Agriculture can have negative effects on biodiversity as well. Organic farming is a multifaceted sustainable agriculture set of practices that can have a lower impact on the environment at the small scale. However, in most cases organic farming results in lower yields in terms of production per unit area.

Therefore, widespread adoption of organic agriculture will require additional land to be cleared and water resources extracted to meet the same level of production. It is believed by many that conventional farming systems cause less rich biodiversity than organic systems. Organic farming has shown to have an average 30% higher species richness than conventional farming. Organic systems on average also have 50% more organisms.

Conservation Tillage

Conservation tillage is an alternative tillage method for farming which is more sustainable for the soil and surrounding ecosystem. This is done by allowing the residue of the previous harvest's crops to remain in the soil before tilling for the next crop. Conservation tillage has shown to improve many things such as soil moisture retention, and reduce erosion. Some disadvantages are the fact that more expensive equipment is needed for this process, more pesticides will need to be used, and the positive effects take a long time to be visible. The barriers of instantiating a conservation tillage policy are that farmers are reluctant to change their methods, and would protest a more expensive, and time consuming method of tillage than the conventional one they are used to.

Other specific methods include: permaculture; and biodynamic agriculture which incorporates a spiritual element.

Questions

Short Answer Type Questions

1. What is green revolution?
2. When does green revolution initiated?
3. Expand GMO.
4. What is residue-free agriculture?
5. Define organic farming.

Essay Type Questions

1. What is modern agriculture and what are its benefits?
2. Write down various components of modern agriculture?
3. How does modern agriculture differ from traditional agriculture?
4. What are the effects of modern agriculture on aquatic ecosystem?
5. Write down the long-term effects of modern agriculture on environment.

7

Desertification

What are Deserts?

When you hear the word 'desert', what picture does it bring to mind? Perhaps you think of a very dry place, a place without people or plants living there. In fact, a deserted place. Perhaps you also think of miles and miles of sand dunes, and a scorching sun burning down from a cloudless sky. One-fifth of the world's land surface is desert. A desert is technically defined as an area which has on average less than 250 mm of rain per year. However, annual rainfall could be as low as 20 mm, leading to periods of intense dryness which often last for several years followed by periods of heavy rainfall and flooding. Surprisingly, more people die in deserts each year from drowning than from thirst.

One of the most famous deserts, the Sahara in North Africa, is very hot and very dry. The Sahara holds the world record for being the hottest place on Earth, with a maximum recorded temperature of 58°C in the shade. Yet at night the desert becomes very cold, with temperatures often falling below freezing.

There are four types of desert landscape. The first, and our most common vision of what a desert should look like, is the sand desert or erg. About one-third of all desert landscapes are like this. The sand is blown into hills or dunes by the wind. These shift over time so that if the movements of the dunes of a sand desert over several years were filmed and then speeded up so that the film lasted only a few minutes, the movements of the desert would look like ripples in the water. It is for this reason that ergs are sometimes called "sand seas".

Desertification is a type of land degradation in which a relatively dry area of land becomes increasingly arid, typically losing its bodies of water as well as vegetation and wildlife. It is caused by a variety of factors, such as through climate change (particularly the current global warming) and through the overexploitation of soil through human activity. When deserts appear automatically over the natural course of a planet's life cycle, then it can be called a natural phenomenon; however, when deserts emerge due to the rampant and unchecked depletion of nutrients in soil that are essential for it to remain arable, then a virtual "soil death" can be spoken of, which traces its cause back to human overexploitation. Desertification is a significant global ecological and environmental problem.

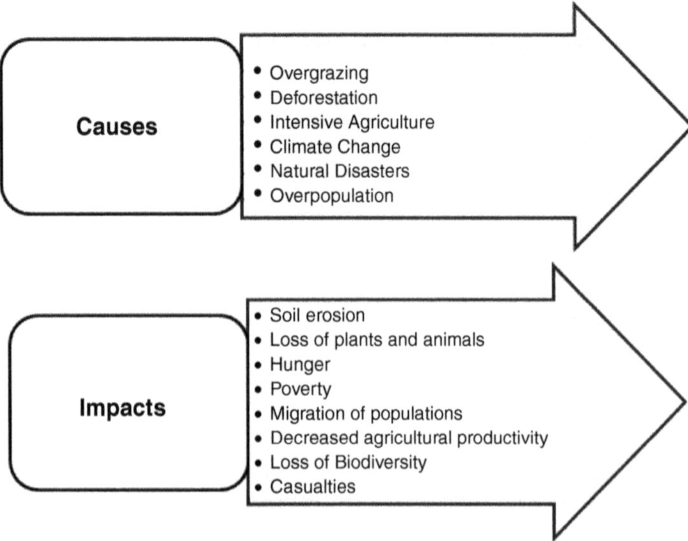

Figure 7.1. Causes and impacts of Desertification

History

The world's most noted deserts have been formed by natural processes interacting over long intervals of time. During most of these times, deserts have grown and shrunk independent of human activities. Paleo-deserts are large sand seas now inactive because they are stabilized by vegetation, some extending beyond the present margins of core deserts, such as the Sahara, the largest hot desert.

Desertification has played a significant role in human history, contributing to the collapse of several large empires, such as Carthage, Greece, and the Roman Empire, as well as causing displacement of local populations. Historical evidence shows that the serious and extensive land deterioration occurring several centuries ago in arid regions had three epicenters: the Mediterranean, the Mesopotamian Valley, and the Loess Plateau of China, where population was dense.

Process of Desertification

Desertification is a process of continuous, gradual ecosystem degradation, during which plants and animals, and geological resources such as water and soil, are stressed beyond their ability to adjust to changing conditions. Because desertification occurs gradually, and the processes responsible for it are understood, it can often be avoided by planning or reversed before irreparable damage occurs.

The physical characteristics of land undergoing desertification include progressive loss of mature, stabilizing vegetation from the ecosystem, or loss of agricultural crop cover during periods of drought or economic infeasibility, and a resulting loss of unconsolidated topsoil. This process is called deflation.

Erosion by wind and water then winnows the fine-grained silt and clay particles from the soil; dramatic dust storms were essentially composed of blowing topsoil. Continued irrigation of decertified land increases soil salinity, and contaminates groundwater, but does little to reverse the loss of productivity. Finally, ongoing wind and water erosion leads to development of gullies and sand dunes across the deflated land surface. The forces causing these physical changes to occur may be divided into natural, human or cultural, and administrative causes.

Among the natural forces are wind and water erosion of soil, long-term changes in rainfall patterns, and other changes in climatic conditions. The role of drought is variable and related in part to its duration; a prolonged drought accompanied by poor land management may be devastating, while a shorter drought might not have lasting consequences. As such, drought thus stresses the ecosystem without necessarily degrading it permanently. Rainfall similarly plays a variable role that depends on its duration, the seasonal pattern of its occurrence, and its spatial distribution. The list of human or cultural influences on desertification includes vegetation loss by overgrazing, depletion of groundwater, surface runoff of rainwater, frequent burning, deforestation, the influence of invasive non-native species, physical compaction of the soil by livestock and vehicles, and damage by strip-mining.

Desertification caused by human influences has a long historical record. Administrative influences contributing to desertification include encouragement of the widespread cultivation of a single crop for export, particularly if irrigation is required, and the concentration of dense human populations in arid lands. Poor economic conditions also contribute to degradation of croplands. During crisis, farmers confront with bankruptcy and a decade-long drought, and they left millions of acres of ploughed, bare cropland unplanted. According to the 1934 Yearbook of Agriculture, "Approximately 35 million acres of formerly cultivated land have essentially been destroyed for crop production.... 100 million acres now in crops have lost all or most of the topsoil; 125 million acres of land now in crops are rapidly losing topsoil."

Considering these factors together, desertification can be viewed as a process of interwoven natural, human, and economic forces causing continuous degradation over time. Therefore, ecosystem and agricultural degradation caused by desertification must be confronted from scientific, social and economic angles. Fortunately, scientists believe that severe desertification, which renders the land irreclaimable, is rare. Most desertified areas can be ecologically reclaimed or restored to agronomic productivity, if socioeconomic and cultural factors permit restoration.

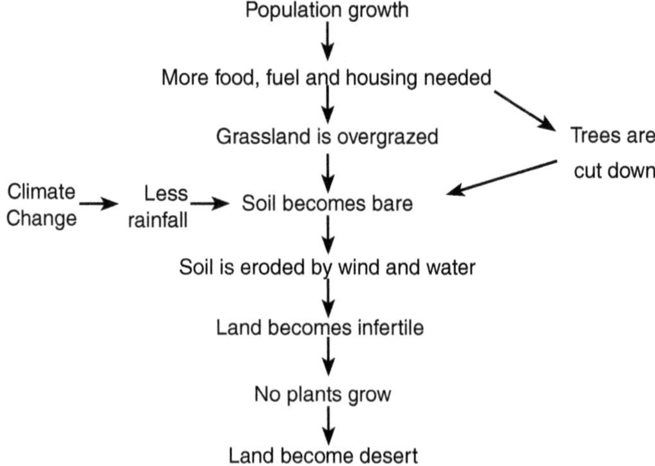

Figure 7.2. Process of Desertification

Causes of Desertification

1. **Overgrazing:** Animal grazing is a huge problem for many areas that are starting to become desert biomes. If there are too many animals that are overgrazing in certain spots, it makes it difficult for the plants to grow back, which hurts the biome and makes it lose its former green glory.

2. **Loss of vegetation:** The immediate cause is the loss of most vegetation. This is driven by a number of factors, alone or in combination, such as drought, climatic shifts, tillage for agriculture, overgrazing and deforestation for fuel or construction materials. Vegetation plays a major role in determining the biological composition of the soil. Studies have shown that, in many environments, the rate of erosion and runoff decreases exponentially with increased vegetation cover. Unprotected, dry soil surfaces blow away with the wind or are washed away by flash floods, leaving infertile lower soil layers that bake in the sun and become an unproductive hardpan.

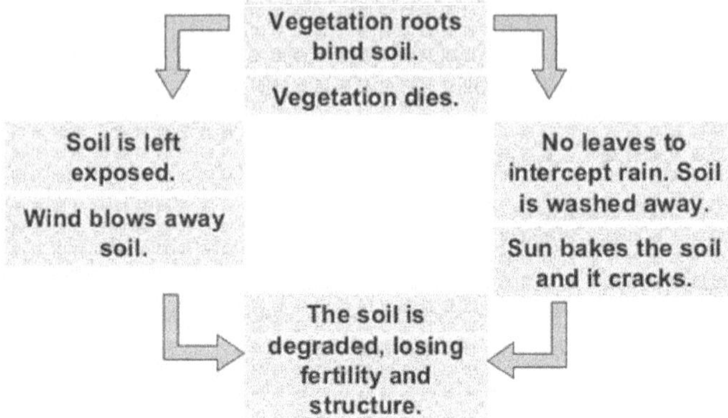

Figure 7.3. Effect of desertification on vegetation

3. **Deforestation:** When people are looking to move into an area, or they need

trees in order to make houses and do other tasks, then they are contributing to the problems related to desertification. Without the plants around, the rest of the biome cannot thrive.

4. **Farming practices:** Some farmers do not know to use the land effectively. They may essentially strip the land of everything that it has before moving on to another plot of land. By stripping the soil of its nutrients, desertification becomes more and more of a reality for the area that is being used for farming.
5. **Urbanization and other types of land development:** As mentioned above, development can cause people to go through and kill the plant life. It can also cause issues with the soil due to chemicals and other things that may harm the ground. As areas become more urbanized, there are less places for plants to grow, thus causing desertification.
6. **Climate change:** Climate change plays a huge role in desertification. As the days get warmer and periods of drought become more frequent, desertification becomes more and more eminent. Unless climate change is slowed down, huge areas of land will become desert, some of those areas may even become inhabitable as time goes on.
7. **Stripping the land for resources:** If an area of land has natural resources like natural gas, oil, minerals, people will come in and mine it or take it out. This usually strips the soil of nutrients, which in turn kills the plant life, which in turn starts the process toward becoming a desert biome as time goes on.
8. **Natural disasters:** There are some cases where the land gets damaged because of natural disasters, including drought. In those cases, there isn't a lot that people can do except work to try and help rehabilitate the land after it has already been damaged by nature.
9. **Poverty and overpopulation**: At least 90% of the inhabitants of dry lands live in developing countries, where they also suffer from poor economic and social conditions. This situation is exacerbated by land degradation because of the reduction in productivity, the precariousness of living conditions and the difficulty of access to resources and opportunities.

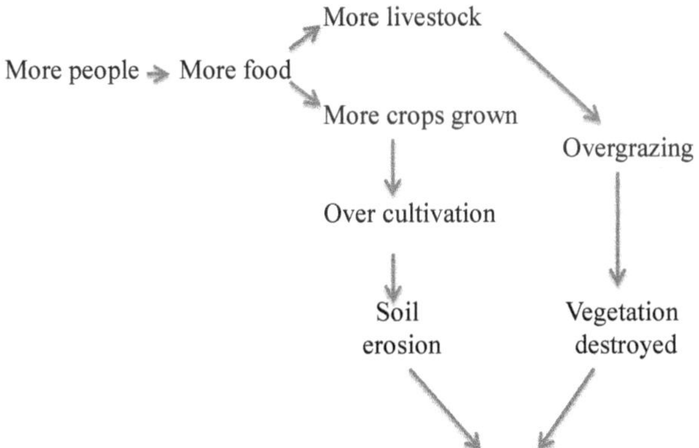

Figure 7.4. Effect of population growth on desertification

Effects of Desertification

1. **Farming:** Farming becomes impossible if an area becomes desert because it can't support substantial crops there without special technologies. This can cost a lot of money to try and do, so many farmers will have to sell their land and leave the desert areas.
2. **Hunger:** With farms in these areas, the food that those farms produce will become much scarcer, and the people who live in those areas will be a lot more likely to deal with hunger problems. Animals will also go hunger, which will cause even more of a food shortage.
3. **Flooding**: Without the plant life in an area, flooding is a lot more eminent. Not all deserts are dry, those that are wet could experience a lot of flooding because there is nothing to stop the water from gathering and going all over the place. Flooding can also negatively affect the water supply.
4. **Poor water quality:** If an area becomes a desert, the water quality is going to become a lot worse than it would have been otherwise. This is because the plant life plays a significant role in keeping the water clean and clear and without its presence, it becomes a lot more difficult to keep the water clean.
5. **Overpopulation:** When areas start to become desert, animals and people will go to other areas where they can actually thrive. This causes crowding and overpopulation, which will, in long run, end up continuing the cycle of desertification that started this whole thing anyway.
6. **Poverty:** All the issues that we have talked about above can lead to poverty if it is not kept in check. Without food and water, it becomes harder for people to thrive and they take a lot of time to try and get the things that they need.

Solutions for Desertification

1. **Policy changes related to how people can farm:** In countries where policy change will actually be enforced on those in the country, policy change related to how often people can farm and how much they can farm on certain areas could be put into place to help reduce the problems that are often associated with farming and desertification.
2. **Policy changes to other types of land use:** If people are using land to get natural resources or they are developing it for people to live on, then the policies that govern them should be ones that will help the land to thrive instead of allowing them to harm the land further.
3. **Education:** In developing countries, education is an incredibly important tool that needs to be utilized in order to help people to understand the best way to use land that they are farming on. By educating them about sustainable practices, more land will be saved from becoming desert.
4. **Technology advances:** In some cases, it is difficult to try and prevent desertification from happening. In those cases, there needs to be research and advancements in technology that push the limits of what we currently know. Advancements could help us find more ways to prevent the issue from becoming epidemic.

5. **Putting together rehabilitation efforts:** There are some ways that we can go back and rehabilitate the land that we have already pushed into desertification; it just takes some investment of time and money. By putting these together, we can prevent the issue from becoming even more widespread in the areas that have already been affected.
6. **Sustainable practices to prevent desertification from happening:** There are plenty of sustainable practices that can be applied to those acts that may be causing desertification. By adding these to what we should be doing with land, we can ensure that we don't turn the entire world into a desert.

Desertification is a huge problem that needs to be addressed accordingly, and if we take the time to do it now, we can prevent other problems from happening with it in the future. By taking that critical look at desertification, we have the tools that we need in order to get through the processes effectively.

United Nation Convention to Combat Desertification (UNCCD)

UNCCD was established in 1994. It is the sole legally binding international agreement linking environment of development to sustainable land management. The convention addresses specifically the arid, semi-arid and dry sub-humid areas, known as drylands.

It was aclopted is Paris, France on 17 June 1994. It has 197 parties, making it near universal in reach.

For the publication of convention, 2006 was declared "International year of Desert of Desertification".

Questions

Short Answer Type Questions

1. Define desertification
2. What is average rainfall per year in deserts?
3. Why deserts are known as sand seas?
4. Name one Paleo-desert.

Essay Type Questions

1. What is desertification? Where does it happen?
2. Can desertification be reversed?
3. What types of plants can be planted to combat desertification?
4. Is desertification linked to climate change?
5. What is the United Nations convention to combat desertification?
6. What are the major causes of desertification?
7. Explain the solutions and preventive measures for desertification.

8

Wetlands and their Conservation

Areas of marsh, fen, peat land or water, whether natural or artificial, permanent or temporary with stagnant or flowing water, freshwater or brackish, including the areas of marine water, the depth of which does not exceed 6 meters is known as wetland. These are ecotones or transitional zones between the permanently aquatic and dry terrestrial ecosystem. These are areas inundated and saturated by surface or groundwater at a frequency and duration sufficient to support vegetation typically adapted in life in saturated soil conditions.

The primary factor that distinguishes wetlands from other land forms or water bodies is the characteristic vegetation of aquatic plants, adapted to the unique hydric soil. Wetlands play a number of roles in the environment, principally water purification, flood control, carbon sink and shoreline stability. Wetlands are also considered the most biologically diverse of all ecosystems, serving as home to a wide range of plant and animal life.

Characteristics of Wetlands

Wetlands vary widely due to local and regional differences in topography, hydrology, vegetation, and other factors, including human involvement.

1. Depth of wetlands should not exceed 6 meters.
2. Soil is mostly hydric, supporting hydrophytes vegetation.
3. Vegetation is specifically adapted to reducing conditions, e.g. Water Lilly.
4. Wetlands are ecotones and thus rich in species diversity.
5. Also known as supermarkets because these are most productive ecosystems of the world comparable to rain forests.

6. Wetlands cover approximately 6% of total land surface.
7. Carbon is stored in plant community and soil instead of releasing them to the atmosphere as carbon dioxide, thus help in moderating global climatic conditions.

Types of Wetlands

1. **Marine Wetlands:** These include coastal lagoons, rocky shore and coral reefs.
2. **Esturine Wetlands:** These include deltas, tidal marshes and mangrove swamps.
3. **Riverine Wetlands:** Places along the rivers and streams.
4. **Lacustrine Wetlands:** These wetlands are associated with the lakes.

Benefits of Wetlands

Wetlands are species rich habitats performing valuable ecosystem services such as flood production, water quality enhancement, and food chain support and carbon sequestration. Although in past people thought wetland as mucky, mosquito infested wasteland, they now recognize their importance, as wetlands are vital to ecological balance of the earth.

1. **Agricultural production:** Wetlands cycle both sediments and nutrients balancing terrestrial and aquatic ecosystems. A natural function of wetland vegetation is the up-take and storage of nutrients found in the surrounding soil and water. For example, well managed rice paddy systems produce not only rice but also co-benefits from rice-associated biodiversity.
2. **Flood peak reduction:** The floodplains of major rivers act as natural storage reservoirs, enabling excess water to spread out over a wide area, which reduces its depth and speed. Wetlands close to the headwaters of streams and rivers can slow down rainwater runoff and spring snowmelt so that it doesn't run straight off the land into water courses. This can help prevent sudden, damaging floods downstream.
3. **Recreational and educational opportunities:** Wetlands provide endless opportunities for popular recreational activities, such as hiking, boating, hunting, fishing, trapping and bird watching.
4. **Socio-economic benefits:** Wetland systems naturally produce an array of vegetation and other ecological products that can harvested for personal and commercial use. The most significant of these is fish which have all or part of their life-cycle occurs within a wetland system. Fresh and saltwater fish are the main source of protein for one billion people and comprise 15% of an additional two billion people's diets. In addition, fish generate a fishing industry that provides 80% of the income and employment to residents in developing countries.
5. **Soil erosion protection:** Vegetated wetlands along the shores of lakes and rivers can protect against erosion caused by waves along the shorelines during floods and storms. Wetlands plants are important because they can

absorb much of energy of the surface waters and bind soil and deposited sediments in their root system.

6. **Water storage:** Although wetlands have often been referred to as natural sponges that soak up water, they actually function more like natural tubs, storing either flood waters that overflow riverbanks or surface water that collects in isolated depressions. As flood waters recede, the water is released slowly from the wetland soils.

7. **Water quality improvement:** Many wetland systems possess bio-filters, hydrophytes, and organisms that in addition to nutrient up-take abilities have the capacity to remove toxic substances that have come from pesticides, industrial discharges, and mining activities. The up-take occurs through most parts of the plant including the stems, roots, and leaves. Floating plants can absorb and filter heavy metals. Water hyacinth (*Eichhornia crassipes*), duckweed (*Lemna*) and water fern (*Azolla*) store iron and copper commonly found in wastewater.

8. **Groundwater replenishment**: Wetland systems are directly linked to groundwater and a crucial regulator of both the quantity and quality of water found below the ground. Wetland systems that are made of permeable sediments like limestone or occur in areas with highly variable and fluctuating water tables especially have a role in groundwater replenishment or water recharge. Sediments that are porous allow water to filter down through the soil and overlying rock into aquifers which are the source of 95% of the world's drinking water. Wetlands can also act as recharge areas when the surrounding water table is low and as a discharge zone when it is too high.

9. **Sediment accretion:** The ability of wetland systems to store nutrients and trap sediment is highly efficient and effective but each system has a threshold. An overabundance of nutrient input from fertilizer run-off, sewage effluent, or non-point pollution will cause eutrophication. Upstream erosion from deforestation can overwhelm wetlands making them shrink in size and see dramatic biodiversity loss through excessive sedimentation load. The capacity of wetland vegetation to store heavy metals is affected by water flow, number of hectares (acres), climate, and type of plant.

10. **Reservoirs of biodiversity:** Wetland system's rich biodiversity is becoming a focal point due to the high number of species present in wetlands, the small global geographic area of wetlands, the number of species which are endemic to wetlands, and the high productivity of wetland systems. Thousands of animal species, 20,000 of them vertebrates, are living in wetland systems. The discovery rate of freshwater fish is at 200 new species per year.

Major Threats

1. **Excessive hunting:** Hunting remains one of the more popular activities on wetlands. Many wetlands are being destroyed for hunting of some important animal species.

2. **Excessive tourism:** Wetland ecosystem has been exploited extensively by the tourists. They pollute the wetlands and even hurt the plant and animal species present over there.
3. **Fisheries:** Due to high nutrient content in wetlands, a large number of fish species are found over there. But when such ecosystems are used as a source of fisheries, they lose their natural regeneration capacity which leads to loss in natural fish population.
4. **Fires:** As wetlands are known as ecotones or transitional zones between terrestrial and aquatic ecosystem, so whenever there is a forest fire in the forest, the terrestrial part of the wetland is destroyed.
5. **Grazing:** The terrestrial portion of the wetlands is destroyed if it is allowed for overgrazing by the cattle. Grass is mainly affected because they are eaten up by the cattle before they could regenerate again.
6. **Increasing nutrients:** Increasing nutrients here means run-off from the agricultural fields which leads to eutrophication. This results in increased productivity initially, but subsequently dissolved oxygen is consumed by the dominating plant species and all other organisms are deprived of it. This ultimately destroys the whole ecosystem.
7. **Pollution due to dumping:** Sometimes the nearby industries and commercial sites, dump their waste into the wetlands. Therefore, a huge amount of toxic wastes get entered the wetland and affect all the organisms there.
8. **Plants and animal pest:** Natural pest also affects the plant and animal species of the wetlands.
9. **Urbanization:** Urbanization contributes to direct habitat loss, modifies hydrological and sedimentation regimes and alters dynamics of nutrient and chemical pollutants. This leads to species dominance and effect ecosystem productivity.
10. **Water diversion:** Water diversions aren't without side effects. Reintroducing sediment will alter the coastal wetlands decreasing the shrimp, fishery and oyster farming.

Conservation of Wetlands

1. **Protection laws and government initiatives:** Proper implementation of laws is very important for the conservation of wetlands. Government initiatives can help to undertake intensive conservation measures and lay down policies for the proper management of wetlands.
2. **Planning and monitoring:** A wetland system needs to be monitored over time to time in order to assess whether it is functioning at an ecologically sustainable level or whether it is becoming degraded. Degraded wetlands will suffer a loss in water quality, a high number

of threatened and endangered species, and poor soil conditions. Due to the large size of wetlands, mapping is an effective tool to monitor wetlands. There are many remote sensing methods that can be used to map wetlands. Remote-sensing technology permits the acquisition of timely digital data on a repetitive basis. This repeat coverage allows wetlands, as well as the adjacent land-cover and land-use types, to be monitored seasonally and/or annually. Using digital data provides a standardized data-collection procedure and an opportunity for data integration within a geographic information system

3. **Controlling pollution**: Due to increase in pollution in environment, wetlands are also affected. Pollution in wetlands is a growing concern, affecting drinking water sources and biological diversity. Drainage and run-off from fertilized crops and pesticides used in industry introduce nitrogen and phosphorus nutrients and other toxins like mercury to water sources. All these pollutants are harmful to various life forms living in the wetlands; therefore pollution should be controlled so that the species that live in wetlands can survive.

Ramsar Convention

Ramsar convention was enacted in Iran on 2nd February 1971 for the conservation of wetlands. It is an intergovernmental treaty which provides the framework for national action and international co-operation for the conservation and wise use of wetlands and their resources.

The primary purposes of the treaty are to list wetlands of international importance and to promote their wise use, with the ultimate goal of preserving the world's wetlands. Methods include restricting access to the majority portion of wetland areas, as well as educating the public to combat the misconception that wetlands are wastelands.

Under the Ramsar international wetland conservation treaty, wetlands are defined as follows:

Article 1.1: "Wetlands are areas of marsh, fen, peat land or water, whether natural or artificial, permanent or temporary, with water that is static or flowing, fresh, brackish or salt, including areas of marine water the depth of which at low tide does not exceed six meters."

Article 2.1: "Wetlands may incorporate riparian and coastal zones adjacent to the wetlands, and islands or bodies of marine water deeper than six meters at low tide lying within the wetlands."

India has 27 designated Ramsar sites covering 1056871 hectares of land including Chilika Lake in Odisha, Chandertaal Lake, HP, etc. The Convention works closely with five International Organization Partners. These are: Birdlife International, the IUCN, the International Water Management Institute, Wetlands International and the World Wide Fund for Nature.

Questions

Short Answer Type Questions

1. Define wetlands.
2. When Ramsar convention came into action?
3. How many Ramsar sites are there in India and all over the world?
4. When is wetland day celebrated?

Essay Type Questions

1. Write down the characteristics of wetlands.
2. What were the objectives of Ramsar convention?
3. What are the different types of wetlands?
4. Write down the importance of wetlands.
5. What is wetland loss and degradation?
6. What are the threats to wetlands?
7. Which is the regulating body governing wetlands in India?
8. Explain the conservation strategies to protect wetlands.

9

Eutrophication

Eutrophication

Eutrophication (from Greek word eutrophos, "well-nourished"), or **hypertrophication**, occurs when a body of water becomes overly enriched with minerals and nutrients that induce excessive growth of plants and algae. This process may result in oxygen depletion of the water body. One example is the "bloom" or great increase of phytoplankton in a water body as a response to increased levels of nutrients. Eutrophication is mainly induced by the discharge of nitrate or phosphate-containing detergents, fertilizers, or sewage into an aquatic system.

Mechanism of Eutrophication

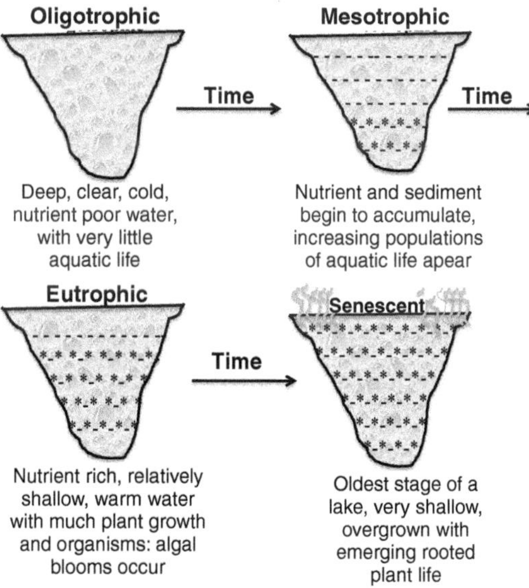

Figure 9.1. Process of Eutrophication

Eutrophication arises from the oversupply of nutrients, which leads to overgrowth of plants and algae. After such organisms die, the bacterial degradation of their biomass consumes the oxygen in the water, thereby creating the state of hypoxia.

According to Ullmann's Encyclopedia, "the primary limiting factor for eutrophication is phosphate." The availability of phosphorus generally promotes excessive plant growth and decay, favoring simple algae and plankton over other more complicated plants, and causes a severe reduction in water quality. Phosphorus is a necessary nutrient for plants to live, and is the limiting factor for plant growth in many freshwater ecosystems. Phosphate adheres tightly to soil, so it is mainly transported by erosion. Once translocated to lakes, the extraction of phosphate into water is slow, hence the difficulty of reversing the effects of eutrophication. However, numerous literature reports are there, that nitrogen is the primary limiting nutrient for the accumulation of algal biomass.

The sources of these excess phosphates are phosphates in detergent, industrial/ domestic run-offs, and fertilizers. With the phasing out of phosphate-containing detergents in the 1970s, industrial/domestic run-off and agriculture have emerged as the dominant contributors to eutrophication.

Types of Eutrophication

1. Cultural Eutrophication

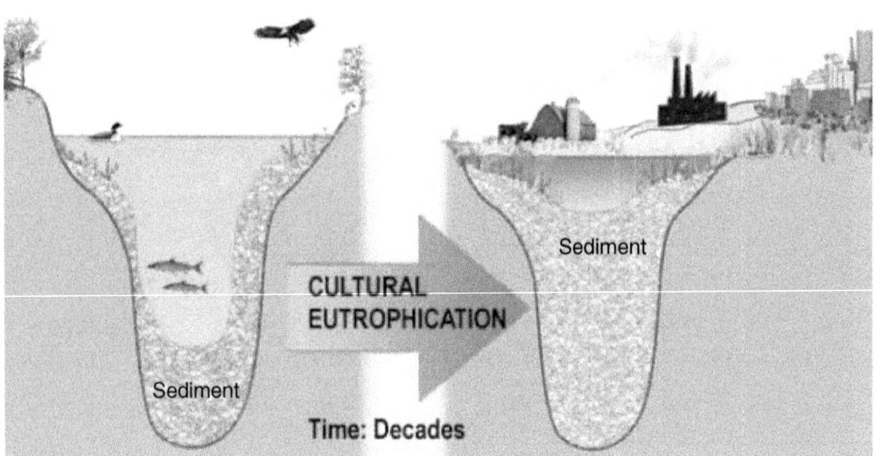

Figure 9.2. Process of cultural eutrophication

Cultural eutrophication is the process that speeds up natural eutrophication because of human activity. Due to clearing of land and building of towns and cities, land runoff is accelerated and more nutrients such as phosphates and nitrate are supplied to lakes and rivers, and then to coastal estuaries and bays. Extra nutrients are also supplied by treatment plants, golf courses, fertilizers, farms, as well as untreated sewage in many countries.

2. Natural Eutrophication

Although eutrophication is commonly caused by human activities, it can also be a natural process, particularly in lakes. Eutrophy occurs in many lakes in temperate grasslands. Paleolimnologists now recognize that climate change, geology, and other external influences are critical in regulating the natural productivity of lakes. Some lakes also demonstrate the reverse process (meiotrophication), becoming less nutrient rich with time. The main difference between natural and anthropogenic eutrophication is that the natural process is very slow, occurring on geological time scales.

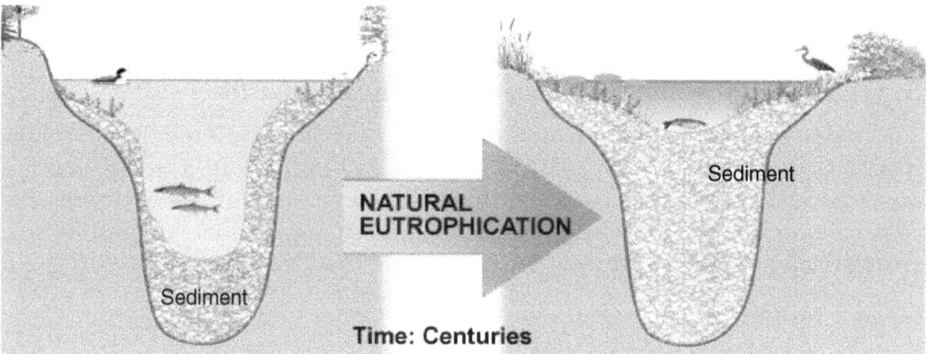

Figure 9.3. Process of Natural Eutrophication

Sources of High Nutrient Run-off Eutrophication

In order to prevent eutrophication from occurring, specific sources that contribute to nutrient loading must be identified. There are two common sources of nutrients and organic matter: point and nonpoint sources.

1. Point sources

Point sources are directly attributable to one influence. In point sources the nutrient waste travels directly from source to water. Point sources are relatively easy to regulate.

- Wastewater effluent (municipal and industrial)
- Runoff and leachate from waste disposal systems
- Runoff and infiltration from animal feedlots
- Runoff from mines, oil fields, unsewered industrial sites
- Overflows of combined storm and sanitary sewers
- Runoff from construction sites less than 20,000 m^2 (220,000 ft^2)
- Untreated sewage

2. Non-point sources

Non-point source pollution (also known as 'diffuse' or 'runoff' pollution) is that which comes from ill-defined and diffuse sources. Non-point sources are difficult to regulate and usually vary spatially and temporally (with season, precipitation, and other irregular events).

- Runoff from agriculture due to fertilizers and pesticides /irrigation
- Runoff from pasture and range
- Urban runoff from unsewered areas
- Septic tank leachate
- Runoff from construction sites >20,000 m^2
- Runoff from abandoned mines
- Atmospheric deposition over a water surface

3. Other Land Activities Generating Contaminants

It has been shown that nitrogen transport is correlated with various indices of human activity in watersheds, including the amount of development. Ploughing in agriculture and development are activities that contribute most to nutrient loading. There are three reasons that nonpoint sources are especially troublesome:

(a) Soil retention

Nutrients from human activities tend to accumulate in soils and remain there for years. It has been shown that the amount of phosphorus lost to surface waters increases linearly with the amount of phosphorus in the soil. Thus much of the nutrient loading in soil eventually makes its way to water. Nitrogen, similarly, has a turnover time of decades.

(b) Runoff to surface water and leaching to groundwater

Nutrients from human activities tend to travel from land to either surface or groundwater. Nitrogen in particular is removed through storm drains, sewage pipes, and other forms of surface runoff. Nutrient losses in runoff and leachate are often associated with agriculture. Modern agriculture often involves the application of nutrients onto fields in order to maximise production. However, farmers frequently apply more nutrients than are taken up by crops or pastures. Regulations aimed at minimising nutrient exports from agriculture are typically far less stringent than those placed on sewage treatment plants and other point source polluters. It should be also noted that lakes within forested land are also under surface runoff influences. Runoff can wash out the mineral nitrogen and phosphorus from detritus and in consequence supply the water bodies leading to slow, natural eutrophication.

(c) Atmospheric deposition

Nitrogen is released into the air because of ammonia volatilization and nitrous oxide production. The combustion of fossil fuels is a large human-initiated contributor to atmospheric nitrogen pollution. Atmospheric nitrogen reaches the ground by two different processes, the first being wet deposition such as rain or snow, and the second being dry deposition which is particles and gases found in the air. Atmospheric deposition (e.g., in the form of acid rain) can also affect nutrient concentration in water, especially in highly industrialized regions.

4. Other Causes

Any factor that causes increased nutrient concentrations can potentially lead to eutrophication. In modeling eutrophication, the rate of water renewal plays a critical role; stagnant water is allowed to collect more nutrients than bodies with replenished water supplies. It has also been shown that the drying of wetlands causes an increase in nutrient concentration and subsequent eutrophication blooms.

Effects of Eutrophication

1. Effects on Lakes and Rivers

When algae die, they decompose and the nutrients contained in that organic matter are converted into inorganic form by microorganisms. This decomposition process consumes oxygen, which reduces the concentration of dissolved oxygen. The depleted oxygen levels in turn may lead to fish kills and a range of other effects reducing biodiversity. Nutrients may become concentrated in an anoxic zone and may only be made available again during autumn turn-over or in conditions of turbulent flow.

Enhanced growth of aquatic vegetation or phytoplankton and algal blooms disrupts normal functioning of the ecosystem, causing a variety of problems such as a lack of oxygen needed for fish and shellfish to survive. The water becomes cloudy, typically coloured a shade of green, yellow, brown, or red. Eutrophication also decreases the value of rivers, lakes and aesthetic enjoyment. Health problems can occur where eutrophic conditions interfere with drinking water treatment.

Human activities can accelerate the rate at which nutrients enter ecosystems. Runoff from agriculture and development, pollution from septic systems and sewers, sewage sludge spreading, and other human-related activities increase the flow of both inorganic nutrients and organic substances into ecosystems. Elevated levels of atmospheric compounds of nitrogen can increase nitrogen availability. Phosphorus is often regarded as the main culprit in cases of eutrophication in lakes subjected to "point source" pollution from sewage pipes. The concentration of algae and the trophic state of lakes correspond well to phosphorus levels in water.

2. Effects on Ocean Waters

Eutrophication is a common phenomenon in coastal waters. In contrast to freshwater systems, nitrogen is more commonly the key limiting nutrient of marine waters; thus, nitrogen levels have greater importance to understanding eutrophication problems in salt water. Estuaries tend to be naturally eutrophic because land-derived nutrients are concentrated where run-off enters a confined channel. Upwelling in coastal systems also promotes increased productivity by conveying deep, nutrient-rich waters to the surface, where the nutrients can be assimilated by algae.

The World Resources Institute has identified 375 hypoxic coastal zones in the world, concentrated in coastal areas in Western Europe, the Eastern and Southern coasts of the US, and East Asia, particularly Japan.

In addition to runoff from land, atmospheric fixed nitrogen can enter the open ocean. A study in 2008 found that this could account for around one-third of the ocean's external (non-recycled) nitrogen supply, and up to 3% of the annual new marine biological production. It has been suggested that accumulating reactive nitrogen in the environment may prove as serious as putting carbon dioxide in the atmosphere.

3. Effects on Terrestrial Ecosystems

Terrestrial ecosystems are subject to similarly adverse impacts from eutrophication. Increased nitrates in soil are frequently undesirable for plants. Many terrestrial plant species are endangered as a result of soil eutrophication, such as the majority of orchid species in Europe. Meadows, forests, and bogs are characterized by low nutrient content and slowly growing species adapted to those levels, so they can be overgrown by faster growing and more competitive species.

Chemical forms of nitrogen are most often of concern with regard to eutrophication, because plants have high nitrogen requirements so that additions of nitrogen compounds will stimulate plant growth. Nitrogen is not readily available in soil because N_2, a gaseous form of nitrogen, is very stable and unavailable directly to higher plants. Terrestrial ecosystems rely on microbial nitrogen fixation to convert N_2 into other forms such as nitrates. However, there is a limit to how much nitrogen can be utilized.

Ecosystems receiving more nitrogen than the plants require are called nitrogen-saturated. Saturated terrestrial ecosystems then can contribute both inorganic and organic nitrogen to freshwater, coastal, and marine eutrophication, where nitrogen is also typically a limiting nutrient. This is also the case with increased levels of phosphorus. However, because phosphorus is generally much less soluble than nitrogen, it is leached from the soil at a much slower rate than nitrogen. Consequently, phosphorus is much more important as a limiting nutrient in aquatic systems.

4. Ecological Effects

Eutrophication was recognized as a water pollution problem in European and North American lakes and reservoirs in the mid-20th century. Since then, it has become more widespread. Surveys showed that 54% of lakes in Asia are eutrophic; in Europe, 53%; in North America, 48%; in South America, 41%; and in Africa, 28%. In South Africa, a study by the CSIR using remote sensing has shown more than 60% of the dams surveyed were eutrophic. Some South African scientists believe that this figure might be higher with the main source being dysfunctional sewage works that produce more than 4 billion liters a day of untreated, or at best partially treated, sewage effluent that discharges into rivers and dams.

Many ecological effects can arise from stimulating primary production, but there are three particularly troubling ecological impacts: decreased biodiversity, changes in species composition and dominance, and toxicity effects.

1. Increased biomass of phytoplankton
2. Toxic or inedible phytoplankton species

3. Increases in blooms of gelatinous zooplankton
4. Increased biomass of benthic and epiphytic algae
5. Changes in macrophyte species composition and biomass
6. Decreases in water transparency (increased turbidity)
7. Colour, smell, and water treatment problems
8. Dissolved oxygen depletion
9. Increased incidences of fish kills
10. Loss of desirable fish species
11. Reductions in harvestable fish and shellfish
12. Decreases in perceived aesthetic value of the water body

5. Decreased Biodiversity

When an ecosystem experiences an increase in nutrients, primary producers reap the benefits first. In aquatic ecosystems, species such as algae experience a population increase (called an algal bloom). Algal blooms limit the sunlight available to bottom-dwelling organisms and cause wide swings in the amount of dissolved oxygen in the water. Oxygen is required by all aerobically respiring plants and animals and it is replenished in daylight by photosynthesizing plants and algae. Under eutrophic conditions, dissolved oxygen greatly increases during the day, but is greatly reduced after dark by the respiring algae and by microorganisms that feed on the increasing mass of dead algae. When dissolved oxygen levels decline to hypoxic levels, fish and other marine animals suffocate. As a result, creatures such as fish, shrimp, and especially immobile bottom dwellers die off. In extreme cases, anaerobic conditions ensue; promoting growth of bacteria such as *Clostridium botulinum* that produces toxins deadly to birds and mammals. Zones where this occurs are known as dead zones.

6. New Species Invasion

Eutrophication may cause competitive release by making abundant a normally limiting nutrient. This process causes shifts in the species composition of ecosystems. For instance, an increase in nitrogen might allow new, competitive species to invade and out-compete original inhabitant species. This has been shown to occur in New England salt marshes. In Europe and Asia, the common carp frequently lives in naturally Eutrophic or Hypereutrophic areas, and is adapted to living in such conditions. The eutrophication of areas outside its natural range partially explain the fish's success in colonising these areas after being introduced.

7. Toxicity

Some algal blooms, otherwise called "nuisance algae" or "harmful algal blooms", are toxic to plants and animals. Toxic compounds they produce can make their way up the food chain, resulting in animal mortality. Freshwater algal blooms can pose a threat to livestock. When the algae die or are eaten, neuro- and hepatotoxins are released which can kill animals and may pose a threat to humans. An example of algal toxins working their way into humans is the case of shellfish poisoning. Biotoxins created during algal blooms are taken up by shellfish (mussels, oysters), leading to these human foods acquiring the toxicity and poisoning humans. Examples include paralytic, neurotoxic, and diarrhoetic shellfish poisoning. Other marine animals can be vectors for such toxins, as in the case of ciguatera, where it is typically a predator fish that accumulates the toxin and then poisons humans.

Prevention

Eutrophication poses a problem not only to ecosystems, but to humans as well. Reducing eutrophication should be a key concern when considering future policy, and a sustainable solution for everyone, including farmers and ranchers, seems feasible. While eutrophication does pose problems, humans should be aware that natural runoff (which causes algal blooms in the wild) is common in ecosystems and should thus not reverse nutrient concentrations beyond normal levels. Clean up measures have been mostly, but not completely, successful. Finnish phosphorus removal measures started in the mid-1970s and have targeted rivers and lakes polluted by industrial and municipal discharges. These efforts have had a 90% removal efficiency. Still, some targeted point sources did not show a decrease in runoff despite reduction efforts.

1. Shellfish in estuaries: unique solutions

One proposed solution to eutrophication in estuaries is to restore shellfish populations, such as oysters and mussels. Oyster reefs remove nitrogen from the water column and filter out suspended solids, subsequently reducing the likelihood or extent of harmful algal blooms or anoxic conditions. Filter feeding activity is considered beneficial to water quality by controlling phytoplankton density and sequestering nutrients, which can be removed from the system through shellfish harvest, buried in the sediments, or lost through denitrification. Foundational work toward the idea of improving marine water quality through shellfish cultivation was conducted by Odd Lindahl *et al.*, using mussels in Sweden. In the United States, shellfish restoration projects have been conducted on the East, West and Gulf coasts.

2. Minimizing nonpoint pollution

Nonpoint pollution is the most difficult source of nutrients to manage. The literature suggests, though, that when these sources are controlled, eutrophication decreases. The following steps are recommended to minimize the amount of pollution that can enter aquatic ecosystems from ambiguous sources.

3. Riparian buffer zones

Studies show that intercepting nonpoint pollution between the source and the water is a successful means of prevention. Riparian buffer zones are interfaces between a flowing body of water and land, and have been created near waterways in an attempt to filter pollutants; sediment and nutrients are deposited here instead of in water. Creating buffer zones near farms and roads is another possible way to prevent nutrients from traveling too far. Still, studies have shown that the effects of atmospheric nitrogen pollution can reach far past the buffer zone. This suggests that the most effective means of prevention is from the primary source.

4. Nitrogen Testing and Modeling

Soil Nitrogen Testing (N-Testing) is a technique that helps farmers optimize the amount of fertilizer applied to crops. By testing fields with this method, farmers saw a decrease in fertilizer application costs, a decrease in nitrogen lost to surrounding sources, or both. By testing the soil and modeling the bare minimum amount of fertilizer are needed, farmers reap economic benefits while reducing pollution.

5. Organic Farming

There has been a study that found that organically fertilized fields "significantly reduce harmful nitrate leaching" over conventionally fertilized fields. However, a more recent study found that eutrophication impacts are in some cases higher from organic production than they are from conventional production.

6. Dredging

Dredging helps in reduction of soil contamination that was caused by sewage sludge (like fecal sludge), wilted plants, or chemical spill. But this may lead to siltation of the water bodies.

7. Bokashi

Bokashi is a modern method of eutrophication prevention to decrease smell and toxic materials. Bokashi is Japanese for "shading off" or "gradation." It derives from the practice of Japanese farmers centuries ago of covering food waste with rich, local soil that contained the microorganisms that would ferment the waste. After a few weeks, they would bury the waste. Bokashi balls used in water pollution are made up of soil (either clay or loam or mixture of both), water, molasses, and antimicrobial solution; antimicrobial solution is used to kill germs (either bacteria or fungi), that cause foul odor, and algae.

Prevention policy

Laws regulating the discharge and treatment of sewage have led to dramatic nutrient reductions to surrounding ecosystems, but it is generally agreed that a policy regulating agricultural use of fertilizer and animal waste must be imposed. In Japan the amount of nitrogen produced by livestock is adequate to serve the fertilizer needs for the agriculture industry. Thus, it is not unreasonable to command livestock owners to clean up animal waste, which when left stagnant will leach into ground water.

Policy concerning the prevention and reduction of eutrophication can be broken down into four sectors: technologies, public participation, economic instruments, and cooperation. The term technology is used loosely, referring to a more widespread use of existing methods rather than an appropriation of new technologies. As mentioned before, nonpoint sources of pollution are the primary contributors to eutrophication, and their effects can be easily minimized through common agricultural practices. Reducing the amount of pollutants that reach a watershed can be achieved through the protection of its forest cover, reducing the amount of erosion leeching into a watershed.

Also, through the efficient, controlled use of land using sustainable agricultural practices to minimize land degradation, the amount of soil runoff and nitrogen-based fertilizers reaching a watershed can be reduced. Waste disposal technology constitutes another factor in eutrophication prevention. Because a major contributor to the nonpoint source nutrient loading of water bodies is untreated domestic sewage, so it is necessary to provide treatment facilities to highly urbanized areas, particularly those in underdeveloped nations, in which treatment of domestic waste water is a scarcity. The technology to safely and efficiently reuse waste water, both from domestic and industrial sources, should be a primary concern for policy regarding eutrophication.

The role of the public is a major factor for the effective prevention of eutrophication. In order for a policy to have any effect, the public must be aware of their contribution to the problem, and ways in which they can reduce their effects. Programs instituted to promote participation in the recycling and elimination of wastes, as well as education on the issue of rational water use are necessary to protect water quality within urbanized areas and adjacent water bodies.

Economic instruments, "which include, among others, property rights, water markets, fiscal and financial instruments, charge systems and liability systems, are gradually becoming a substantive component of the management tool set used for pollution control and water allocation decisions." Incentives for those who practice clean, renewable, water management technologies are an effective means of encouraging pollution prevention. By internalizing the costs associated with the negative effects on the environment, governments are able to encourage a cleaner water management.

Because a body of water can have an effect on a range of people reaching far beyond that of the watershed, cooperation between different organizations is necessary to prevent the intrusion of contaminants that can lead to eutrophication. Agencies ranging from state governments to those of water resource management and non-governmental organizations, going as low as the local population, are responsible for preventing eutrophication of water bodies. In the United States, the most well known inter-state effort to prevent eutrophication is the Chesapeake Bay.

Questions

Short Answer Type Questions

1. Define eutrophication
2. Name the types of eutrophication.
3. Name one oligotrophic lake.
4. Name one mesotropic lake.
5. Name one eutropic lake.

Essay Type Questions

1. Explain the causes of eutrophication.
2. Write down the various stage of eutrophication.
3. What is the effect of eutrophication on aquatic ecosystems?
4. Describe the preventive measures take to avoid the eutrophication of water bodies.

10

Environmental Effect of Minning

The environmental impact of mining includes erosion, formation of sinkholes, loss of biodiversity, and contamination of soil, groundwater and surface water by chemicals from mining processes. Besides creating environmental damage, the contamination resulting from leakage of chemicals also affects the health of the local population. Mining companies in some countries are required to follow environmental and rehabilitation codes, ensuring the area mined is returned close to its original state.

Erosion of exposed hillsides, mine dumps, tailings dams and resultant siltation of drainages, creeks and rivers can significantly impact the surrounding areas. In wilderness areas mining may cause destruction and disturbance of ecosystems and habitats, and in areas of farming, it may disturb or destroy productive grazing and croplands. In urbanized environments mining may produce noise pollution, dust pollution and visual pollution.

1) Water Pollution

Mining can have adverse effects on surrounding surface and groundwater if protective measures are not taken. The result can be unnaturally high concentrations of some chemicals, such as arsenic, sulfuric acid, and mercury over a significant area of surface or subsurface. Runoff of mere soil or rock debris, although non-toxic— also devastates the surrounding vegetation. The dumping of the runoff in surface waters or in forests is the worst option here. Submarine tailings disposal is regarded as a better option (if the soil is pumped to a great depth). Mere land storage and refilling of the mine after it has been depleted is even better, if no forests need to be cleared for the storage of the debris. There is potential for massive contamination of the area surrounding mines due to the various chemicals used in the mining process as well as the potentially damaging compounds and metals removed from the ground with the ore.

Large amounts of water produced from mine drainage, mine cooling, aqueous extraction and other mining processes increases the potential for these chemicals to contaminate ground and surface water. In well-regulated mines, hydrologists and geologists take careful measurements of water and soil to exclude any type of water contamination that could be caused by the mine's operations. The reducing or eliminating of environmental degradation is enforced in modern American mining by federal and state law, by restricting operators to meet standards for protecting surface and groundwater from contamination. This is best done through the use of non-toxic extraction processes as bioleaching.

2) Acid Rock Drainage

Sub-surface mining often progresses below the water table, so water must be constantly pumped out of the mine in order to prevent flooding. When a mine is abandoned, the pumping ceases, and water floods the mine. This introduction of water is the initial step in most acid rock drainage situations.

Acid rock drainage occurs naturally within some environments as part of the rock weathering process but is exacerbated by large-scale earth disturbances characteristic of mining and other large construction activities, usually within rocks containing an abundance of sulfide minerals. Areas where the earth has been disturbed (e.g. construction sites, subdivisions, and transportation corridors) may create acid rock drainage. In many localities, the liquid that drains from coal stocks, coal handling facilities, coal washeries, and coal waste tips can be highly acidic, and in such cases it is treated as acid mine drainage (AMD).

The same type of chemical reactions and processes may occur through the disturbance of acid sulfate soils formed under coastal or estuarine conditions after the last major sea level rise, and constitutes a similar environmental hazard.

The five principal technologies used to monitor and control water flow at mine sites are diversion systems, containment ponds, and groundwater pumping systems, subsurface drainage systems, and subsurface barriers. In the case of AMD, contaminated water is generally pumped to a treatment facility that neutralizes the contaminants.

3) Addition of Heavy Metals

Water in the mine containing dissolved heavy metals such as lead and cadmium, leaks into local groundwater and contaminate it. Long-term storage of tailings and dust can lead to additional problems, as they can be easily blown off site by wind.

Concentrations of heavy metals are known to decrease with distance from the mine, and effects on biodiversity follow the same pattern. Impacts can vary greatly depending on mobility and bioavailability of the contaminant: less-mobile molecules will stay inert in the environment while highly mobile molecules will easily move into another compartment or be taken up by organisms. For example, speciation of metals in sediments could modify their bioavailability, and thus their toxicity for aquatic organisms.

Effects of Mining

1. Effects on Biodiversity

The implantation of a mine is a major habitat modification, and smaller perturbations occur on a larger scale than exploitation site, mine-waste residuals contamination of the environment for example. Adverse effects can be observed long after the end of the mine activity. Destruction or drastic modification of the original site and anthropogenic substances release can have major impact on biodiversity in the area. Destruction of the habitat is the main component of biodiversity losses, but direct poisoning caused by mine-extracted material, and indirect poisoning through food and water, can also affect animals, vegetation and microorganisms.

Habitat modification such as pH and temperature modification disturb communities in the area. Endemic species are especially sensitive, since they need very specific environmental conditions. Destruction or slight modification of their habitat puts them at the risk of extinction. Habitats can be damaged when there is not enough terrestrial product as well as by non-chemicals products, such as large rocks from the mines that are discarded in the surrounding landscape with no concern for impacts on natural habitat.

Biomagnifications plays an important role in polluted habitats: mining impacts on biodiversity should be, assuming that concentration levels are not high enough to directly kill exposed organisms, greater on the species on top of the food chain because of this phenomenon.

Adverse mining effects on biodiversity depend a great extent on the nature of the contaminant, the level of concentration at which it can be found in the environment, and the nature of the ecosystem itself. Some species are quite resistant to anthropogenic disturbances, while some others will completely disappear from the contaminated zone. Time alone does not seem to allow the habitat to recover completely from the contamination. Remediation takes time, and in most cases will not enable the recovery of the original diversity present before the mining activity took place.

2. Effects on Aquatic Organisms

The mining industry can impact aquatic biodiversity through different ways. Direct poisoning is the first one, and risks are higher when contaminants are mobile in the sediment or bioavailable in the water. Mine drainage can modify water pH, and it is hard to differentiate direct impact on organisms from impacts caused by pH changes. Effects can nonetheless be observed and proved to be caused by pH modifications. Contaminants can also affect aquatic organisms through physical effects like streams with high concentrations of suspended sediment limit light, thus diminishing algal biomass. Metal oxide deposition can limit biomass by coating algae or their substrate, thereby preventing colonization.

Factors that impact communities in acid mine drainage sites vary temporarily and seasonally: temperature, rainfall, pH, salinisation and metal quantity all display variations on the long term, and can heavily affect communities. Changes in pH or temperature can affect metal solubility, and thereby the bioavailable quantity that directly impact organisms. Moreover, contamination persists over time: ninety years after a pyrite mine closure, water pH was still very low and microorganism's populations consisted mainly of acidophilic bacteria.

3. Effect on Microorganisms

Algae communities are less diverse in acidic water containing high zinc concentration, and mine drainage stress decrease their primary production. Diatoms community is greatly modified by any chemical addiction to water body. pH phytoplankton assemblage, and high metal concentration diminishes the abundance of planktonic species. Some diatom species may, however, grow in high-metal-concentration sediments. In sediments close to the surface, cysts suffer from corrosion and heavy coating. In much polluted conditions, total algal biomass is quite low, and the planktonic diatom community missing. In case of functional complementarity, however, it is possible that phytoplankton and zooplankton mass remains stable.

4. Effect on Macroorganisms

Water insect and crustacean communities are modified around a mine, resulting in a low trophic completeness and community being dominated by predators. However, biodiversity of macro invertebrates can remain high, if sensitive species are replaced with tolerant ones. When diversity is on the contrary reduced, there is sometimes no effect of stream contamination on abundance or biomass, suggesting that tolerant species fulfilling the same function take the place of sensible species in polluted sites. pH diminution in addition to elevated metal concentration can also have adverse effects on macroinvertebrates' behaviour, showing that direct toxicity is not the only issue. Fishes are also affected by pH, temperature variations and chemical concentrations.

5. Effect on Terrestrial Organisms

(i) Effect on Vegetation

Soils' texture and water content can be greatly modified in disturbed sites, leading to plants communities changes in the area. Most of the plants have a low concentration tolerance for metals in the soil, but sensitivity differs among species. Grass diversity and total cover is less affected by high contaminant concentration than forbs and shrubs. Mines waste-material rejects or traces due to mining activity can be found in the vicinity of the mine, sometimes pretty far away from the source. Established plants cannot move away from perturbations, and will eventually die if their habitat is contaminated by heavy metals or metalloids at concentration too elevated for their physiology. Some species are more resistant and will survive these levels, and some non-native species that can tolerate these concentrations in the soil, will migrate in the mine surrounding lands to occupy the ecological niche.

Plants can be affected through direct poisoning, for example, arsenic soil content reduces bryophyte diversity. Soil acidification through pH diminution by chemical contamination can also lead to a diminished species number. Contaminants can modify or disturb microorganisms, thus modifying nutrient availability, causing a loss of vegetation in the area. Some tree roots avoid the deeper soil layer in order to avoid the contaminated zone, and thus miss anchorage and might be uprooted by the wind when their height and shoot weight increase. In general, root exploration is reduced in contaminated areas compared to non-polluted ones. Even in reclaimed habitats, plant species diversity is lower than in undisturbed areas.

Cultivated crops might be a problem near mines. Most crops can grow on weakly contaminated sites, but yield is generally lower than it would have been in regular growing conditions. Plants also tend to accumulate heavy metals in their aerial organs, possibly leading to human intake through fruits and vegetables. Regular consumptions might lead to health problems caused by long-term metal exposure. Cigarettes made from tobacco growing on contaminated sites may have adverse effects on human population, as tobacco tends to accumulate cadmium and zinc in its leaves.

(ii) Effect on Animals

Habitat destruction is one of the main issues of mining activity. Huge areas of natural habitat are destroyed during mine construction and exploitation, forcing animals to leave the site.

Animals can be poisoned directly by mine products and residuals. Bioaccumulation in the plants or the smaller organisms eaten by them, can also lead to poisoning: horses, goats and sheep are exposed in certain areas to potentially toxic concentration of copper and lead in grass. There are fewer number of ants species in soil containing high copper levels, in the vicinity of a copper mine. If fewer ants are found, chances are great that other organisms living in the surrounding landscape are strongly affected as well by these high copper levels, since ants are a good environmental indicator: they live directly in the soil and are thus pretty sensitive to environmental disruptions.

(iii) Effect on Microorganisms

Because of their size, microorganisms are extremely sensitive to environmental modification, such as modified pH, temperature changes or chemicals concentration. For example, the presence of arsenic and antimony in soils led to a diminution in total soil bacteria. Moreover, as in water, a small change in the soil pH can provoke the remobilization of contaminants, in addition to the direct impact on pH-sensitive organisms.

Microorganisms have a wide variety of genes among their total population, so there is a greater chance of survival of the species due to the existence of resistance or tolerance genes in some colonies, as long as modifications are not too extreme. Nevertheless, survival in these conditions will imply a big loss of gene diversity, resulting in reduced potential adaptations to subsequent changes. The presence of few developed soil in heavy metal contaminated areas could be a sign of reduced activity by soils micro fauna and micro flora, indicating a reduced number of individuals or reduced activity.

Twenty years after disturbance, even in rehabilitation area, microbial biomass is still greatly reduced compared to undisturbed habitat. Arbuscular mycorrhiza fungi are especially sensitive to the presence of chemicals, and the soil is sometimes so disturbed that they are no longer able to associate with root plants. Some fungi possess, however, contaminant accumulation capacity, soil cleaning capacity by changing the biodisponibility of contaminants, and can protect plants from damages caused by chemicals. Their presence in contaminated sites could prevent loss of biodiversity due to mine-waste contamination, or allow bioremediation, that is, the removal of undesired chemicals from contaminated soils. On the contrary, some microbes can deteriorate the environment: which mean elevated SO_4 in the water can also increase microbial production of hydrogen sulfide, a toxin for many aquatic plants and organisms.

6. Effects of Mine Pollution on Humans

Humans are also affected by mining. There are many diseases that can come from the pollutants that are released into the air and water during the mining process. For example, during smelting operations enormous quantities of air pollutants, such as the suspended particulate matter, SO_x, arsenic particles and cadmium, are emitted. Metals are usually emitted into the air as particulates.

There are also many occupational health hazards. Most of the miners suffer from various respiratory and skin diseases. Miners working in different types of mines suffer from asbestosis, silicosis, or black lung disease. Humans are also affected by the occurrence of landslides and floods.

Coal Mining

Deforestation

With open cast mining the overburden, which may be covered in forest, must be removed before the mining can commence. Although the deforestation due to mining may be small compared to the total amount it may lead to species extinction if there is a high level of local endemism.

Sand Mining

Sand mining and gravel mining creates large pits and fissures in the earth's surface. At times, mining can extend so deeply that it affects groundwater, springs, underground wells, and the water table.

Mitigation

To ensure completion of reclamation, or restoring mine land for future use, many governments and regulatory authorities around the world require that mining companies post a bond to be held in escrow until productivity of reclaimed land has been convincingly demonstrated, although if cleanup procedures are more expensive than the size of the bond, the bond may simply be abandoned. Since 1978 the mining industry has reclaimed more than 2 million acres (8,000 km^2) of land in the United States alone. This reclaimed land has renewed vegetation and wildlife in previous mining lands and can even be used for farming and ranching.

Questions

Short Answer Type Questions

1. Name some important non-metallic minerals.
2. Name the main mineral fuels.
3. Define mining.

Essay Type Questions

1. Explain the types of minerals and their uses.
2. Describe different types mining processes.
3. How does mining affect the land resources?
4. Write the effects of mining on aquatic ecosystem.
5. Which types of mining affects the atmosphere to a greater extent and how does mining contribute to air pollution?
6. Which group of human population is severely affected by mining?
7. How does the extraction of mineral resources affect humans?

11

Wasteland Reclamation

Wastelands are lands which are unproductive, unfit for cultivation, grazing and other economic uses due to rough terrain and eroded soils. The lands which are waterlogged and saline are also termed as wastelands. The loss of fertility followed by erosion also leads to the conversion of marginal forest lands into wastelands. It can be defined as 'land that is economically unproductive and presently not in use due to various deteriorating factors'. For example degraded forest lands, mine spoils, waterlogged lands, marshy and saline lands, steep slopes, etc.

Land is an important resource since it is put to different uses by man. India has a land area of nearly 32.88Xkm^2 which is about 2.4 percent of the world. Agriculture in India covers 51% of the total area, 4% by pasture, 19% by forests and 26% by the wastelands. Loss of vegetation cover leads to loss of soil through erosion, which ultimately creates wastelands. In India out of total land area almost half occur as wastelands and Rajasthan being the state with maximum waste land area.

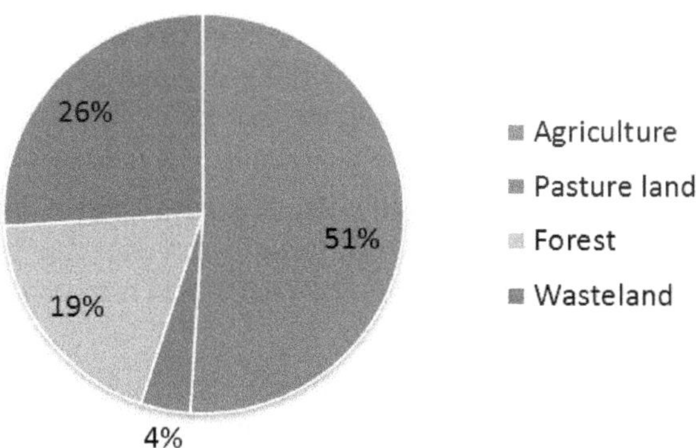

Figure 11.1. Reasons for waste land formation

In the absence of land management policy, geomorphic processes become active due to which soil layers are eroded and transported, making these lands infertile, stony and useless. This is one of pressing problems of one country as loss of soil has already ruined large amounts of cultivable lands. If it remains unchecked it will affect the remaining lands. Hence conservation of soil, protecting the existing cultivable lands and reclaiming the already depleted wastelands is becoming important.

There are many reasons for waste land formation. Like deforestation which leads to soil erosion and the eroded soils exhibit drought tendency. Further, the falling of trees aggravates the lowering of water table and dry conditions. The loss of fertility followed by erosion also leads to the transformation of marginal forest lands into wastelands.

1. **Natural reasons**: the natural forces like volcanic eruptions, storms, earthquakes, floods, salinity and snow covering creates wastelands.
2. **Human reasons**: human activities like mining, bad agricultural practices, deforestation, overgrazing and deposition of solid waste and chemical wastes create wastelands.

Classification of Wastelands

The Wastelands Can be Classified into Two Categories

1. **Barren and uncultivable wastelands:** These lands cannot be brought under cultivation or economic use except at a very high cost, whether they exist as isolated pockets or within cultivated holdings. Such lands are sandy desert, gully land, stony or leached land, lands on hilly slopes, rocky exposures, etc.
2. **Cultivable wastelands:** These lands are not cultivated for five years or more. It consists of lands available for cultivation, but not used for cultivation. Next to fallow lands, cultivable wastelands are important for agricultural purposes because they can be reclaimed through conservational methods for cultivation, grazing or agroforestry.

Types of Waste Land Soil

1. **Acidic soil**: These are the soils which have pH less than 5. The soil acidity develops due to high rainfall which leaches more soluble bases (like Ca^{+2}, Na^+, K^+) downward or acidic parental rock. The acidic soils bear only sparse vegetation.
2. **Saline soil:** These are the soils which have excess of salts dominated by chlorides and sulphates. The saline soils develop due to groundwater or canal water rich in basic salts. The vegetation is poor in these soils. They are known by name reh in UP.
3. **Alkaline or Sodic soil**: These are the soils which contain high exchangeable sodium or magnesium percentage. These soils are unproductive due to waterlogging. The pH ranges from 8.5 to 10.

Environmental Effects of Wasteland Formation

1. Surface runoff and floods.
2. Soil erosion and desertification.
3. Loss of nutrients and land productivity.
4. Soil acidification.
5. Soil salinity
6. Loss of biodiversity
7. Long-term impact on humans which include migration.

Reclamation of Wastelands

Reclamation means to re-claim the wasteland or to use it for productive purpose. It is a process of turning barren, sterile wasteland into something that is fertile and suitable for habitation and cultivation. Need of reclamation includes:

1. Source of income.
2. Constant supply of fuel, fodder and timber for local use.
3. It makes soil fertile by preventing soil erosion and conserving moisture.
4. Ecological balance in area is maintained.
5. Maintain local climate due to reforestation.

The reclamation and development of wasteland has four major ecological objectives:

1. To improve the physical structure and quality of the soil.
2. To improve availability and quality of water.
3. To prevent the shifting of soil, landslides and flooding.
4. To conserve the biological resources of the land for sustainable use.

According to the ease of reclaim ability, wastelands can be categorized into three forms:

1. **Easily reclaimable**: This form of wasteland can be used for agriculture, as it can be easily reclaimed by applying compost to increase the fertility, removing salination by reducing salt content or by using other fertilizers like gypsum, urea and potash.
2. **Reclaimed with some difficulty**: This form of wasteland cannot be used for agriculture only, as the damage to this land is more. Therefore this wasteland can be used for agroforestry, which includes multiple use of the land like growing trees, crops and livestock management.
3. **Reclaimed with extreme difficulty**: In this type of wasteland, the degradation is much more, so cannot be used for agriculture. It can be used for forestry or natural ecosystem. Plant species like Eucalyptus, Poplar, Acacia, etc. can be grown as they grow well in alkaline water deficient soils.

Reclamation of acidic, saline and alkali soil is the process of their conversion into such type of soils which will be fit for cultivation of crops and tree plantations. The various methods employed in reclamation are:

1. Mechanical Methods

1. **Salt removal:** The salts deposited on earth's surface are removed by scrapers.
2. **Salt leaching:** The extra salt in soil is removed by using excess water which pushes the excess of salt downward.
3. **Drainage:** Underground horizontal drainage technique is applied which consists of underlying porous pipes that pick up salt water and drain the same to the outside.
4. **Better irrigation practices:** Like surface irrigation, fountain irrigation, sprinkle irrigation, drip irrigation, etc. are followed.

2. Chemical Methods

The most common chemical method of reducing soil alkalinity is addition of gypsum. It absorbs free sodium and form neutral salt sodium sulphate.

$$\text{Soil (Na)}_2 + CaSO_4 = \text{Soil-Ca} + Na_2SO_4 \text{ (Sodium sulphate)}$$

3. Biological Methods

1. **Afforestation and Reforestation:** Aforestation means growing the forest over cultivable wasteland and reforestation means growing the forest again over the lands where they were existing and was destroyed due to fires, overgrazing, and excessive cutting. Reforestation checks waterlogging, floods, soil erosion and increase productivity of land. National Wasteland Development Board (NWDB) was constituted in 1985 by Govt. of India for reclamation of soil by planting trees.
2. **Providing surface cover:** The easiest way to protect the land surface from soil erosion is of leave crop residue on the land after harvesting.
3. **Mulching:** Here also protective cover of organic matter and plants like stalks, cotton stalks, tobacco stalks, etc. are used which reduce evaporation, help in retaining soil moisture and reduce soil erosion.
4. **Changing ground topography on downhills:** Running water erodes the hill soil and carries the soil along with it. This can be minimized by following alteration in ground topography:
5. **Strip farming:** Different kinds of crops are planted in alternate strip along the contour
6. **Contour ploughing:** In this arrangement, the ploughing of land is done across the hill and not in up and down style.
7. **Terracing:** In this arrangement, the earth is shaped in the form of leveled terraces to hold soil and water. The terrace edges are planted with such plant species which anchor the soil.

8. **Applying green manures and bio-fertilizers:** Application of green manure of plants like *Sesbania sps.* and sunn hemp have been reported to improve salt-affected soils. Blue-green algae also improve salt affected soils.
9. **Selection of high salt tolerant crops:** The crops like Barley, Sugar beet and Date palms are highly salt tolerant and they show good results in terms of yield in salt containing soils.
10. **Ecological succession:** This refers to the natural development or redevelopment of an ecosystem which help in reclaiming the minerally deficient soils of wasteland.

Questions

Short answer type questions

1. Define wasteland.
2. What are the natural reasons of wasteland formation?
3. Write the human reasons of wasteland formation.

Essay type questions

1. Explain the type of wastelands.
2. What are the causes of wasteland formation, explain in detail?
3. Describe the various methods of wasteland reclamation.

12

Deforestation

Deforestation, clearance, or clearing is the removal of a forest or stand of trees where the land is thereafter converted to a non-forest use. Examples of deforestation include conversion of forestland to farms or urban use. The most concentrated deforestation occurs in tropical rainforests. About 30 per cent of Earth's land surface is covered by forests, but deforestation is clearing these essential habitats on a massive scale. The world's rain forests could completely vanish in a hundred years at the current rate of deforestation.

The biggest driver of deforestation is agriculture. Farmers cut forests to provide more room for planting crops or grazing livestock. Often, small farmers will clear a few acres by cutting down trees and burning them in a process known as burn agriculture.

Logging operations, which provide the world's wood and paper products, also cut countless trees each year. Loggers, some of them acting illegally, also build roads to access more and more remote forests, which lead to further deforestation. Forests are also cut as a result of growing urban sprawl as land is developed for dwellings. Not all deforestation is intentional. Some is caused by a combination of human and natural factors like wildfires and subsequent overgrazing, which may prevent the growth of young trees.

The removal of trees without sufficient reforestation has resulted in habitat damage, biodiversity loss, and aridity. It has adverse impacts on bio-sequestration of atmospheric carbon dioxide. Deforestation has also been used in war to deprive the enemy of vital resources and cover for its forces.

Deforestation causes extinction, changes to climatic conditions, desertification, and displacement of populations as observed by current conditions and in the past through the fossil record. More than half of all plant and land animal species in the world live in tropical forests.

Causes of Deforestation

According to the United Nations Framework Convention on Climate Change (UNFCCC) secretariat, the overwhelming direct cause of deforestation is agriculture. Subsistence farming is responsible for 48% of deforestation; commercial agriculture is responsible for 32%; logging is responsible for 14%, and fuel wood removals make up 5%.

Experts do not agree on whether industrial logging is an important contributor to global deforestation. Some argue that poor people are more likely to clear forest because they have no alternatives, others that the poor lack the ability to pay for the materials and labor needed to clear forest. One study found that population increases due to high fertility rates were a primary driver of tropical deforestation in only 8% of cases.

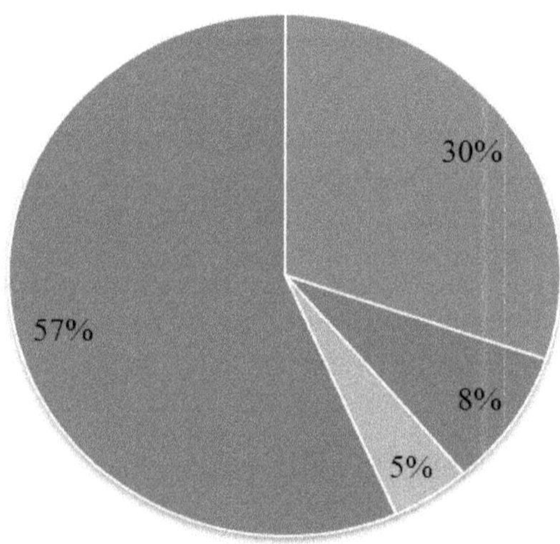

Figure 12.1. Major causes of Deforestation

Other causes of contemporary deforestation may include corruption of government institutions, the inequitable distribution of wealth and power, population growth and overpopulation, and urbanization. Globalization is often viewed as another root cause of deforestation, though there are cases in which the impacts of globalization have promoted localized forest recovery. Major causes of deforestation are discussed below:

1. **Conversion of forest land to agriculture land:** Due to the tremendous increase in population, the demand of food and shelter is also increasing. This increasing demand for food has led to increased agriculture practices which further needs more and more land for cultivation. This ultimately leads to the clearance of forests for agriculture.

2. **Forest fire:** Forest fire is caused by both natural and anthropogenic reasons. Natural causes include-volcanic eruptions and lightening whereas anthropogenic causes are arson, discarded cigarettes, campfires left unattended, burning of debris. Forest fires not only affect the vegetation immediately, but also affect the productivity of the forest in long term.
3. **Unofficial cutting:** Illegal cutting of forest for industrial purpose like paper industry; for timber purpose and mainly for extracting some important medicines is increasing at a very high rate. The plants of medicinal purpose are being fell off and are smuggled, due to which some plant species have become extinct in their home range area.
4. **Overgrazing:** Uncontrolled grazing of the open forests by the cattle reduces the time required by the plants to regenerate. Plants are being eaten up by the animal before they could regenerate and results in death of the plant.
5. **Road and transport development:** For the construction of roads and railway tracks, a large number of trees are felt down and forest land is converted to cemented area and this cemented area could never support plant growth again. Recently at least 54000 mangroves spread across 13.36 ha in Maharashtra will be used for Mumbai-Ahmedabad bullet train project.
6. **Development of industries and township:** Industrialization and urbanization are one of the major contributors for deforestation. Industries require area for the industrial establishment and raw material for which forests are destroyed. Increasing size of cities is interfering with the natural forest ecosystem.
7. **River valley project:** Dam construction needs huge area for the construction of water reservoir. For the construction of reservoir, nearby forest is being cleared and the forest land is submerged in water forever.

Environmental Effects

Deforestation can have a negative impact on the environment. The most dramatic impact is a loss of habitat for millions of species. Eighty per cent of Earth's land animals and plants live in forests, and many cannot survive if deforestation destroy their homes.

Deforestation also drives climate change. Forest soils are moist, but without protection from sun-blocking tree cover, they quickly dry out. Trees also help perpetuate the water cycle by returning water vapor to the atmosphere. Without trees to fill these roles, many former forest lands can quickly become barren deserts. Removing trees deprives the forest of portions of its canopy, which blocks the sun's rays during the day, and holds in heat at night. This disruption leads to more extreme temperature swings that can be harmful to plants and animals.

Trees also play a critical role in absorbing the greenhouse gases that fuel global warming. Fewer forests means larger amounts of gases entering the atmosphere; and increased speed and severity of global warming.

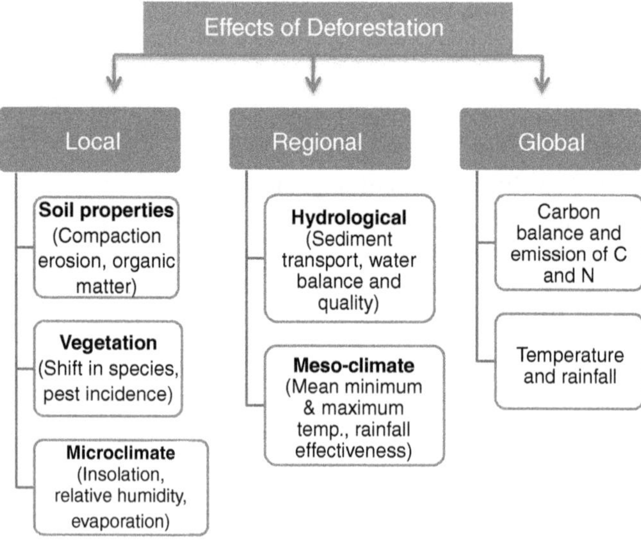

Figure 12.2. Effects of Deforestation

Effects of deforestation can be categorized in two categories:

1. Direct effects
2. Indirect effects

Direct Effects

1. Loss of biodiversity: Deforestation on a human scale results in decline in biodiversity, and on a natural global scale is known to cause the extinction of many species. The removal or destruction of areas of forest cover has resulted in a degraded environment with reduced biodiversity. Forests support biodiversity, providing habitat for wildlife; moreover, forests foster medicinal conservation

2. Natural imbalance: Reducing forest cover reduces the capacity of the soil to perspire. It means the absence of trees can influence the quantity of water on land, in atmosphere or in soil. It effects the ecological cycle.

3. Loss for habitat for animals: Deforestation is the massive cutting down and clearing of trees and plants. Because many animal species that live in the forests are herbivorous, such as giraffes, deer and tapirs, they are forced to leave what once was their home to search for food. As a result, many starve to death. Other times they out of the forest and into areas populated by humans and get hit by automobiles. Meanwhile, carnivorous animals that prey on herbivorous animals find it more difficult to find food and end up starving to death or wandering off into human-populated areas and getting killed. Interrelation of monkeys in the residential areas in one of the greatest example of habital loss.

4. Loss of tribal culture: As large amounts of forests are cleared away, allowing exposed earth to wither and die and the habitats of innumerable species to be destroyed, the indigenous tribes who depend on them to sustain their way of life are also irreparably damaged. Forest is essential for the survival of tribes or *Adivasis*.

It is the means of their livelihood. The ways and the rules of the tribes in the use of the forest are inherently sustainable as forest conservation is in their blood.

5. Decreased rainfall: The water cycle is also affected by deforestation. Trees extract groundwater through their roots and release it into the atmosphere. When part of a forest is removed, the trees no longer transpire this water, resulting in a much drier climate.

6. Soil erosion: Deforestation reduces the content of water in the soil and groundwater as well as atmospheric moisture. The dry soil leads to lower water intake for the trees to extract. Deforestation reduces soil cohesion, so that erosion.

7. Interception: Shrinking forest cover lessens the landscape's capacity to intercept, retain and transpire precipitation. Instead of trapping precipitation, which then percolates to groundwater systems, deforested areas become sources of surface water runoff, which moves much faster than subsurface flows.

Indirect Effects

1. **Increase in environmental pollution:** There are grave consequences for forest destruction. Its biggest disadvantage is in the form of air pollution. The air where there is lack of trees gets polluted. And the problem of air pollution is the highest in the cities. There people suffer from many diseases, especially breathing problems such as asthma.
2. **Increase in greenhouse effect:** Deforestation causes carbon dioxide to linger in the atmosphere. As carbon dioxide accrues, it produces a layer in the atmosphere that traps radiation from the sun. The radiation converts to heat which causes global warming, which is better known as the greenhouse effect. Plants remove carbon in the form of carbon dioxide from the atmosphere during the process of photosynthesis, but release some carbon dioxide back into the atmosphere during normal respiration.
3. **Ozone layer depletion:** The normal environment of the Earth as a result of deforestation has become polluted. It is posing grave danger to the ozone layer, which is necessary for the overall defense of the Earth. Imagine that bad day (May it never come), when the ozone layer disappears.
4. **Increased frequency of landslides and disastrous floods:** Forests return most of the water that falls as precipitation to the atmosphere by transpiration. In contrast, when an area is deforested, almost all precipitation is lost as run-off. That quicker transport of surface water can translate into flash flooding and more localized floods than would occur with the forest cover. Soils are reinforced by the presence of trees, which secure the soil by binding their roots to soil bedrock. Due to deforestation, the removal of trees causes sloped lands to be more susceptible to landslides.
5. **Climate change:** Deforestation has a direct impact on the natural climate change, thereby increasing the global temperature. With the decreasing area of forests, the rain is also becoming irregular. This contributes to 'global warming', which has direct impact on humans.
6. **Decrease in forest productivity:** The forest products industry is a large part of the economy in both developed and developing countries. Short-

term economic gains made by conversion of forest to agriculture, or over-exploitation of wood products, typically leads to loss of long-term income and long-term biological productivity.

Control of Deforestation

The most feasible solution to deforestation is to carefully manage forest resources by eliminating clear-cutting to make sure forest environments remain intact. The cutting that does occur should be balanced by planting young trees to replace older trees felled. The number of new tree plantations is growing each year, but their total still equals a tiny fraction of the Earth's forested land.

1. **Reducing emissions:** In evaluating implications of overall emissions reductions, countries of greatest concern are those categorized as High Forest Cover with High Rates of Deforestation (HFHD) and Low Forest Cover with High Rates of Deforestation (LFHD). Control in the emissions can prevent the destruction of natural forest ecosystems due to pollution.
2. **Payments for conserving forests:** The local people residing near the forests, if conserve the forest in any sense, should be paid for the same. This would encourage them to perform these practices and conserve the forest. This has been made mandatory in many countries.
3. **Land rights:** Transferring rights over land from public domain to its indigenous inhabitants is argued to be a cost effective strategy to conserve forests. This includes the protection of such rights entitled in existing laws, such as India's Forest Rights Act. The transferring of such rights in China, perhaps the largest land reform in modern times, has been argued to have increased forest cover.
4. **Farming:** New methods are being developed to farm more intensively, such as high-yield hybrid crops, greenhouse and hydroponics. These methods are often dependent on chemical inputs to maintain necessary yields. In cyclic agriculture, cattle are grazed on farm land that is resting and rejuvenating. Cyclic agriculture actually increases the fertility of the soil. Intensive farming can also decrease soil nutrients by consuming at an accelerated rate the trace minerals needed for crop growth. The most promising approach, however, is the concept of food forests in permaculture, which consists of agro-forestal systems carefully designed to mimic natural forests, with an emphasis on plant and animal species of interest for food, timber and other uses. These systems have low dependence on fossil fuels and agro-chemicals, are highly self-maintaining, highly productive, and with strong positive impact on soil and water quality, and biodiversity.
5. **Monitoring deforestation:** There are multiple methods that are appropriate and reliable for reducing and monitoring deforestation. One method is the "visual interpretation of aerial photos or satellite imagery that is labor-intensive but does not require high-level training in computer image processing or extensive computational resources". Another method includes hot-spot analysis, using expert opinion or coarse resolution satellite data to identify locations for detailed digital analysis with high resolution satellite

images. Deforestation is typically assessed by quantifying the amount of area deforested, measured at the present time. From an environmental point of view, quantifying the damage and its possible consequences is a more important task, while conservation efforts are more focused on forested land protection and development of land-use alternatives to avoid continued deforestation.

6. **Forest management:** Efforts to stop or slow deforestation have been attempted for many centuries because it has long been known that deforestation can cause environmental damage sufficient in some cases to cause societies to collapse. In the areas where "slash-and-burn" is practiced, switching to "slash-and-char" would prevent the rapid deforestation and subsequent degradation of soils. The biochar thus created, given back to the soil, is not only a durable carbon sequestration method, but it also is an extremely beneficial amendment to the soil. Mixed with biomass it brings the creation of terra preta, one of the richest soils on the planet and the only one known to regenerate itself.
7. **Sustainable practices:** According to the United Nations Food and Agriculture Organization (FAO), "A major condition for the adoption of sustainable forest management is a demand for products that are produced sustainably and consumer willingness to pay for the higher costs entailed. Certification represents a shift from regulatory approaches to market incentives to promote sustainable forest management. By promoting the positive attributes of forest products from sustainably managed forests, certification focuses on the demand side of environmental conservation." Rainforest Rescue argues that the standards of organizations like FSC are too closely connected to industry interests and therefore do not guarantee environmentally and socially responsible forest management. In reality, monitoring systems are inadequate and various cases of fraud have been documented worldwide.
8. **Reforestation:** In many parts of the world, reforestation and aforestation are increasing the area of forested lands. In Western countries, increasing consumer demand for wood products that have been produced and harvested in a sustainable manner is causing forest landowners and forest industries to become increasingly accountable for their forest management and timber harvesting projects.
9. **Forest plantations:** Globally, planted forests increased from 4.1% to 7.0% of the total forest area between 1990 and 2015. Plantation forests made up 280 million ha in 2015, an increase of about 40 million ha in the last ten years. Globally, planted forests consist of about 18% exotic or introduced species while the rest are species native to the country where they are planted. The trees will protect local villages from storm damages and will provide a habitat for local wildlife.

Conclusion: Today, forests are being cut indiscriminately, resulting in seasonal changes, increase in soil heat, ozone layer depletion, etc. Our development process has displaced thousands of people from water, forests and lands. Protecting the forest in this critical situation should not only be our duty but our religion. Let us all take a pledge to stop the indiscriminate harvesting of trees and make an invaluable contribution in saving the forest. If in reality we have to protect the heritage of the forest for the coming generations, then we will have to develop a thinking to save forests for life, not to just derive benefits from them. We will have to learn how to respect the forest, and change our ways of thinking.

Questions

Short Answer Type Questions

1. What is a forest?
2. Define deforestation.
3. Define aforestation.
4. Define reforestation.
5. What is plantation forest?

Essay Type Questions

1. Explain the causes of deforestation.
2. How does deforestation effects the environment?
3. Describe the strategies for forest conservation.
4. Explain the following terms:
   (a) Agroforestry
   (b) Social forestry
   (c) Overgrazing
   (d) Shift cultivation
   (e) Shelter forest

13

Air Pollution

Atmosphere

We all know that earth is a unique planet due to the presence of life. The air is one among the necessary conditions for the existence of life on this planet. The air is a mixture of several gases and it encompasses the earth from all sides. The air surrounding the earth is called the atmosphere. It envelops the earth all round and is held in place by the gravity of the earth. Generally, atmosphere extends up to about 1600 km from the earth's surface. However, 99 % of the total mass of the atmosphere is confined to the height of 32 km from the earth's surface.

Composition of Atmosphere

The atmosphere is composed of a mix of several different gases in differing amounts. The permanent gases whose percentages do not change from day-to-day are nitrogen, oxygen and argon. Nitrogen accounts for 78% of the atmosphere, oxygen 21% and argon 0.9%. Gases like carbon dioxide, nitrous oxides, methane, and ozone are trace gases that account for about a tenth of one per cent of the atmosphere. Water vapor is unique in that its concentration varies from 0-4% of the atmosphere depending on where you are and what time of the day it is.

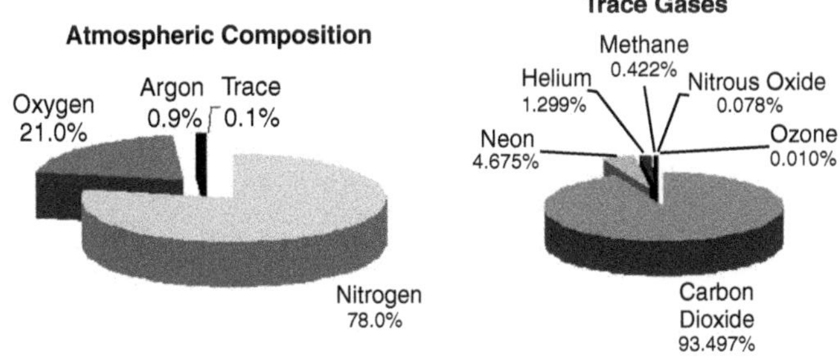

Figure 13.1. Composition of Atmosphere

1. **Troposphere** - 0 to 12 km - contains 75% of the gases in the atmosphere. This is where life exists live and where weather occurs. As height increases, temperature decreases. The temperature drops about 6.5 degrees Celsius for every kilometer above the earth's surface. Tropopause - located at the top of the troposhere. The temperature remains fairly constant here. This layer separates the troposphere from the stratosphere. We find the jet stream here. These are very strong winds that blow eastward.
2. **Stratosphere** - 12 to 50 km - in the lower part of the stratosphere, the temperature remains fairly constant (–60 degrees Celsius). This layer contains the ozone layer. Ozone acts as a shield for in the earth's surface. It absorbs ultraviolet radiation from the sun. This causes a temperature increase in the upper part of the layer.
3. **Mesophere** - 50 to 80 km - in the lower part of the stratosphere, the temperature drops in this layer to about –100 degrees Celsius. This is the coldest region of the atmosphere. This layer protects the earth from meteoroids. They burn up in this area.
4. **Thermosphere** - 80 km and up - Thermosphere means "heat sphere". The temperature is very high in this layer because ultraviolet radiation is turned into heat. Temperatures often reach 2000 degrees Celsius or more. This layer contains:
    1. **Ionosphere** - This is the lower part of the thermosphere. It extends from about 80 to 550 km. Gas particles absorb ultraviolet and X-ray radiation from the sun. The particles of gas become electrically charged (ions). Radio waves are bounced off the ions and reflect waves back to earth. This generally helps radio communication. However, solar flares can increase the number of ions and can interfere with the transmission of some radio waves.
    2. **Exosphere** - The upper part of the thermosphere. It extends from about 550 km for thousands of kilometers. Air is very thin here. This is the area where satellites orbit the earth.

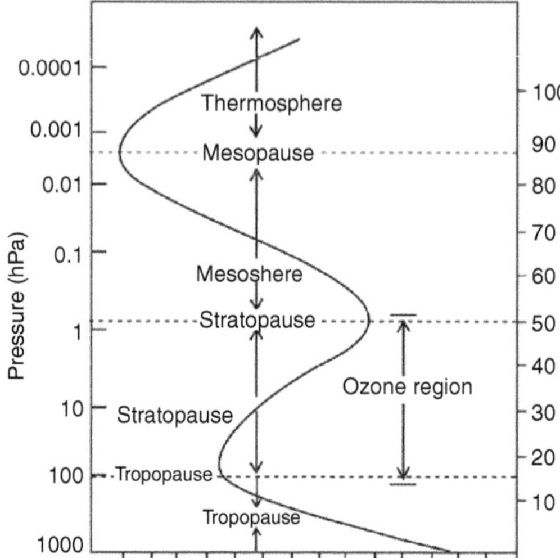

Figure 13.2. Various layers of Atmosphere

Air Pollution

Air pollution may be defined as the presence of any solid, liquid or gaseous substance including noise and radioactive radiation in the atmosphere in such concentration that may be directly and indirectly injurious to humans or other living organisms, plants, property or interferes with the normal environmental processes. A variety of organic compounds are emitted into the atmosphere by natural and human activities. They are so diverse that it is difficult to classify them neatly.

Type and factors of air pollution vary country to country, for instance in Sri Lanka it is particles and CO, in South Africa and India SO_2 from Coal power plants and CO from house hold combustion and forest fires.

Sources of Air pollution

I Natural Sources

Compounds released from volcanic activities such as black smoke, ash, metals, SO_x, CO_x and release of methane from thawing of permafrost regions in the northern hemisphere, wetlands, sanitary landfills. Forest fires and bush fires, dust storm, sea spray and conversion of land use and release of isoprenes and terpenes by forest (precursors of low level ozone).

II Anthropogenic Sources

1) Sources of Energy Generation

This is where CO_x and SO_x and water vapor are released in the atmosphere as large amount of coal, oil, L.P/ natural gas, gasoline and bio-fuels are used in combustion.

2) Transport

This is mobile and most leading source of CO. Combustion in engines is mainly fueled by gas, petrol, diesel, and kerosene. Jet engines of sub sonic long range air craft's are major source of NO_x. Traffic on road is considered as non-point or line source, addition to that harbors and turbine engines of huge ships also emits tons of greenhouse gases and toxic particles in the air.

3) Industry

Most of the industries are directly or indirectly depend on fossil fuel, as they produce CO and CO_2, sulfur hexafluoride and particle matters. Mainly cement industry releases large amount of particle matters in the environment. There is an array of hazardous volatile compounds that are released from paints, electronics, dry cleansing, decreasing agents. Furthermore, utilization of HFC, oxides of nitrogen, PFC and SF_6 produces pollutants.

4) Households

Carbon and soot emission during the cooking by the use of fossil fuels can be considered here. Volatile toxicants such as permethrine compounds from insecticides could contaminate in the air or even food and resulting in the intoxication.

5) Agricultural Practices

Agriculture activities such as use of natural fertilizer release greenhouse gases. Pesticides release persistent organic pollutants (POP). Enteric fermentation in cattle ranching produces greenhouse gases mainly methane. Toxic chemicals found in pesticide and weedicide also reduces the quality of air inhaled.

6) Land Mining, Earth Moving Activity and Quarrying

Process of mining large mineral deposits in the earth accompanied with emission of dust and other chemicals. Blasting, quarrying limestone in cement manufacturing produces dust particles.

7) Construction and Repair Works

Drilling, blasting, transportation, loading and unloading activities often causes dust generation. In addition, there are several nonpoint anthropogenic sources related to dust generation such as welding, painting, auto mobile repairing, etc.

8) Burning of Wastes and Incinerators

This is more severe threat to the environment as it contaminates the atmosphere with persistent organic pollutants (POP) such as dioxins, furans probably major sources are plastics and electronic wastes. In addition, as in normal combustion carbon is emitted as oxides and soot. Wastes are in a vast array such as plastic, electronic wastes, cement dust, industrial chemicals, paper, glass, steel and various derivatives of soil minerals, biological and medicinal wastes, drugs and other chemicals. Incinerators destroys the hazardous effect of any gas or particle and the remaining dust emission could be as smalls as PM10-PM2.5 or lesser, unless right particle filters are used. It also ends up with adverse results.

Types of Pollutants

Primary and Secondary Pollutants

Pollutants can be classified as primary or secondary. Primary are the gases and particles released to the atmosphere and remain in the same form as it is from the source, whereas the secondary are generated by the chemical reaction of primary pollutants in the atmosphere.

Primary pollutants are substances that are directly emitted into the atmosphere from sources. The main primary pollutants known to cause harm in high enough concentrations are the following:

- ✰ Carbon compounds, such as CO, CO_2, CH_4, and VOCs
- ✰ Nitrogen compounds, such as NO, N_2O, and NH_3
- ✰ Sulfur compounds, such as H_2S and SO_2
- ✰ Halogen compounds, such as chlorides, fluorides, and bromides

Particulate Matter (PM or "aerosols"), either in solid or liquid form, which is usually categorized into these groups based on the aerodynamic diameter of the particles:

1. Particles less than 100 microns, which are also called "inhalable" since they can easily enter the nose and mouth.
2. Particles less than 10 microns (PM10, often labeled "fine" in Europe). These particles are also called "thoracic" since they can penetrate deep in the respiratory system.
3. Particles less than 4 microns. These particles are often called "respirable" because they are small enough to pass completely through the respiratory system and enter the bloodstream.
4. Particles less than 2.5 microns (PM 2.5, labeled "fine" in the US). Particles less than 0.1 microns (PM0.1, "ultrafine"). Sulfur compounds were responsible for the traditional wintertime sulfur smog in London in the mid 20th century. These anthropogenic pollutants have sometimes reached lethal concentrations in the atmosphere, such as during the infamous London episode of December 1952.

Secondary pollutants are not directly emitted from sources, but instead form in the atmosphere from primary pollutants (also called "precursors"). The main secondary pollutants known to cause harm in high enough concentrations are the following:

- NO_2 and HNO_3 formed from NO
- Ozone (O_3) formed from photochemical reactions of nitrogen oxides and VOCs
- Sulfuric acid droplets formed from SO_2, and nitric acid droplets formed from NO_2
- Sulfates and nitrates aerosols (e.g., ammonium (bi) sulfate and ammonium nitrate) formed from reactions of sulfuric acid droplets and nitric acid droplets with NH_3, respectively
- Organic aerosols formed from VOCs in gas-to-particle reactions

Major Pollutants of Air

1) Oxides of Nitrogen

Nitric oxide (NO) and nitrogen dioxide (NO_2) are the two most important nitrogen oxide air pollutants. They are frequently lumped together under the designation NO_x. Of the two, NO_2 is the more toxic and irritating compound. Nitric oxide is a principal by-product of combustion processes, arising from the high-temperature reaction between N_2 and O_2 in the combustion air and from the oxidation of organically bound nitrogen in certain fuels such as coal and oil. The oxidation of N_2 by the O_2 in combustion air occurs primarily through the two reactions known as the Zeldovich mechanism.

$$N_2 + O \rightarrow NO + N$$

$$N + O_2 \rightarrow NO + O$$

The first reaction above has a relatively high activation energy, due to the need to break the strong N_2 bond. Because of the high activation energy, the first reaction is the rate-limiting step for NO production, proceeds at a somewhat slower rate than the combustion of the fuel, and is highly temperature sensitive. Nitric oxide formed via this route is referred to as *thermal*-NO_x.

Mobile combustion and fossil-fuel power generation are the two largest anthropogenic sources of NO_x In addition, industrial processes and agricultural operations produce minor quantities. Emissions are generally reported as though the compound being emitted were NO_2. Large concentrations can reduce visibility and increase the risk of acute and chronic respiratory disease.

2) Sulfur Oxides

SO_2 is colorless, but has a suffocating, pungent odor. Sulfur dioxide (SO_2) is formed from the oxidation of sulfur contained in fuel as well as from certain industrial processes that utilize sulfur-containing compounds. Anthropogenic emissions of SO_2 result almost exclusively from stationary point sources. Stationary fuel combustion (primarily utility and industrial) and industrial processes (primarily smelting) are the main SO_2 sources. Stationary fuel combustion includes all boilers, heaters, and furnaces found in utilities, industry, and commercial, institutional and residential establishments. Coal combustion has traditionally been the largest stationary fuel combustion source. The primary source of SO_2 is the combustion of sulfur-containing fuels (e.g., oil and coal). Exposure to SO_2 can cause the irritation of lung tissues and can damage health and materials.

3) Carbon Monoxide

This odorless, colorless gas is formed from the incomplete combustion of fuels. Thus, the largest source of CO today is motor vehicles, combustion of fossil fuel, gas, charcoal and wood, naturally from forest fires and volcanoes. It causes difficulty in breathing as it compete with oxygen by forming carboxyhemoglobin, Asphyxia, damage to heart and nervous system.

4) Carbon Dioxide

CO_2 is considered as an air pollutant as it defined by the clean air act. And it is a greenhouse gas increasing levels of CO_2 causing global warming. CO_2 emission is available from all kind of combustion both natural and man-made.

5) Ground Level Ozone

Tropospheric ("low-level") ozone is a secondary pollutant formed when sunlight causes photochemical reactions involving NO_x and VOCs. Automobiles are the largest source of VOCs necessary for these reactions. Ozone concentrations tend to peak in the afternoon, and can cause eye irritation, aggravation of respiratory diseases, and damage to plants and animals.

6) Total Suspended Particulate Matter (TSP)

Total suspended particles (TSP) are divided in two subcategories – particles smaller than 10 μm in diameter (PM_{10}), and particles smaller than 2.5 μm in diameter ($PM_{2.5}$). Particulate matter (PM) can exist in solid or liquid form, and includes smoke, dust, aerosols, metallic oxides, and pollen. Sources of PM include combustion, factories, construction, demolition, agricultural activities, motor vehicles, and wood burning. Inhalation of enough PM over time increases the risk of chronic respiratory disease.

Lead (Pb): The largest source of Pb in the atmosphere has been from leaded gasoline combustion, but with the gradual elimination worldwide of lead in gasoline, air Pb levels have decreased considerably. Other airborne sources include combustion of solid waste, coal, and oils, emissions from iron and steel production and lead smelters, and tobacco smoke. Exposure to Pb can affect the blood, kidneys, and nervous, immune, cardiovascular, and reproductive systems.

7) Polycyclic Aromatic Hydrocarbons

Polycyclic aromatic hydrocarbons are released from cigarette smoke and stove smoke, can cause lung cancer.

Radon

Released naturally from volcanic eruption. It is a radioactive material which ionizes biological molecules and causes cell disruption and lung cancer.

Dioxin

Dioxin is a toxic gas produced from burning of electronic wastes and plastic materials; it could cause cancer and affect the immune system and leads to developmental reproductive disorders.

Furans

Furans are released during the burning of plastic products such as nylon which contains various harmful compounds.

Peroxyacetylnitrate (PAN)

PAN is formed due to photochemical reaction of NO_x with hydrocarbons in the sunlight. It is a component of photochemical. Smog, Smog is a mixture of air pollutants such as gases and particles. PAN often causes irritation to eye and together with ozone it lowers the lung capacity and increases breathing rate.

8) Asbestos

Asbestos fiber dust released from building material, mines, mills and insulations causes mesothelioma, lung cancer, asbestosis.

9) Arsenic

Found in copper smelters and cigarette smoke and causes in lung cancer.

10) Allergens

House dust, pollen, animal dander causes asthma and rhinitis.

Out comes of Air Pollution

1) Photochemical smog

Photochemical smog was first observed in Los Angeles, U.S.A in the mid-1940s and since then this phenomenon has been detected in most major metropolitan cities of the world. The conditions for the formation of photochemical smog are air stagnation, abundant sunlight, and high concentrations of hydrocarbon and nitrogen oxides in the atmosphere. Smog arises from the photochemical reactions in the lower atmosphere by the interaction of hydrocarbons and nitrogen oxide released by exhausts of automobiles and some stationary sources. This interaction results in a series of complex reactions producing secondary pollutants such as ozone, aldehydes, ketones, and peroxyacyl nitrates.

There is characteristic variation with time of the day in levels of NO, NO_2, hydrocarbons, aldehydes and oxidants under smoggy atmospheric conditions in a city with heavy vehicular traffic. It has been seen that shortly after sunrise, the level of NO in the atmosphere decreases markedly followed quickly by increase in NO_2. NO_2 reacts with sunlight leading to various chain reactions and ultimately to the production of ozone and other oxidants. During the mid-day when the concentration of NO has fallen to a very low level, the levels of aldehydes and oxidants become relatively high. The concentration of total hydrocarbon in the atmosphere reaches maximum in the morning and then decreases during the remaining daylight hours. The typical smog episode occurs in hot, sunny weather under low humidity conditions. The characteristic symptoms are the brown haze in the atmosphere, reduced visibility, eye irritation, respiratory distress and plant damage. Three days of apocalyptic haze and smog turned Delhi into a gas chamber in 2019, due to industrial and vehicular pollution.

The control of photochemical smog may require substantial reduction in NO_x produced in urban areas. At the same time it is necessary to control the release of hydrocarbons from numerous mobile and stationary sources. Catalytic converters are now used to destroy the pollutants in exhaust gases. A reduction catalyst is employed to reduce NO in the exhaust gas and after injection of air an oxidation catalyst is used to oxidise the hydrocarbon and CO.

2) Acid Rain

Acid rain refers to a mixture of deposited material, both wet and dry, coming from the atmosphere containing more than normal amounts of nitric and sulfuric acids. Simply put, it means rain that is acidic in nature due to the presence of certain pollutants in the air due to cars and industrial processes. The term acid rain was coined in 1852 by Scottish chemist Robert Angus Smith, according to the Royal Society of Chemistry, which calls him the "father of acid rain."

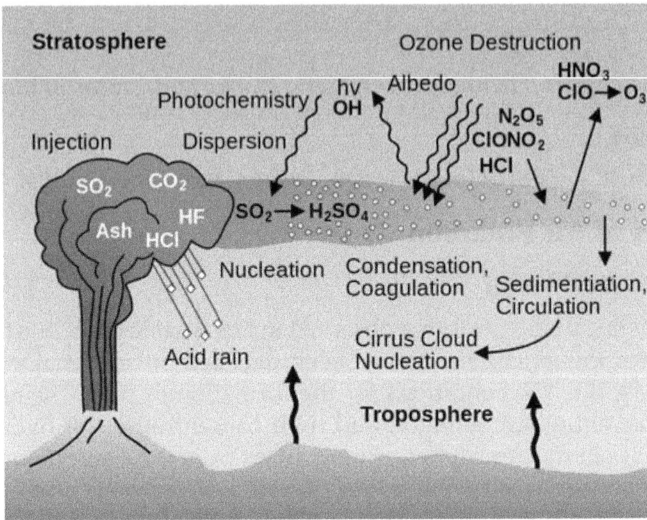

Figure 13.3. Process of Acid Rain

It is easily defined as rain, fog, sleet or snow that has been made acidic by pollutants in the air as a result of fossil fuel and industrial combustions that mostly emits nitrogen oxides (NO_x) and sulfur dioxide (SO_2). Acidity is determined on the basis of the pH level of the water droplets. Normal rainwater is slightly acidic with a pH range of 5.3-6.0, because carbon dioxide and water present in the air react together to form carbonic acid, which is a weak acid. When the pH level of rain water falls below this range, it becomes acid rain.

Effects of acid rain

1) Effect on Aquatic Environment: Acid rain either falls directly on aquatic bodies or gets run off the forests, roads and fields to flow into streams, rivers and lakes. Over a period of time, acids get accumulated in the water and lower the overall pH of the water body. The aquatic plants and animals need a particular pH level of about 4.8 to survive. If the pH level falls below that the conditions become hostile for the survival of aquatic life.

2) Effect on Forests: It makes trees vulnerable to disease, extreme weather, and insects by destroying their leaves, damaging the bark and arresting their growth.

3) Effect on Soil: Acid rain highly impacts on soil chemistry and biology. It means, soil microbes and biological activity as well as soil chemical compositions such as soil pH are damaged or reversed due to the effects of acid rain. The soil needs to maintain an optimum pH level for the continuity of biological activity. When acid rains seep into the soil, it means higher soil pH, which damages or reverses soil biological and chemical activities. Hence, sensitive soil microorganisms that cannot adapt to changes in pH are killed. High soil acidity also denatures enzymes for the soil microbes.

4) Effect on Architecture and Buildings: Acid rain on buildings, especially those constructed with limestone, react with the minerals and corrode them away. This leaves the building weak and susceptible to decay. Modern buildings, cars, airplanes, steel bridges and pipes are all affected by acid rain. Irreplaceable damage can be caused to the old heritage buildings.

5) Effect on Public Health: When in atmosphere, sulfur dioxide and nitrogen oxide gases and their particulate matter derivatives like sulfates and nitrates, degrades visibility and can cause accidents, leading to injuries and deaths. Human health is not directly affected by acid rain because acid rain water is too dilute to cause serious health problems. However, the dry depositions also known as gaseous particulates in the air which in this case are nitrogen oxides and sulfur dioxide can cause serious health problems when inhaled. Intensified levels of acid depositions in dry form in the air can cause lung and heart problems such as bronchitis and asthma.

3) Global Warming

Global warming is the current increase in temperature of the Earth's surface (both land and water) as well as its atmosphere. Average temperatures around the world have risen by 0.75°C (1.4°F) over the last 100 years about two-third of this increase has occurred since 1975. In the past, when the Earth experienced increases in temperature it was the result of natural causes but today it is being caused by the accumulation of greenhouse gases in the atmosphere produced by human activities.

Global warming occurs when carbon dioxide (CO_2) and other air pollutants and greenhouse gases collect in the atmosphere and absorb sunlight and solar radiation that have bounced off the earth's surface. Normally, this radiation would escape into space—but these pollutants, which can last for years to centuries in the atmosphere, trap the heat and cause the planet to get hotter. That's what's known as the greenhouse effect.

- ✰ Melting glaciers, early snowmelt, and severe droughts will cause more dramatic water shortages and increase the risk of wildfires in the American West.
- ✰ Rising sea levels will lead to coastal flooding on the Eastern Seaboard, especially in Florida, and in other areas such as the Gulf of Mexico.
- ✰ Forests, farms, and cities will face troublesome new pests, heat waves, heavy downpours, and increased flooding. All those factors will damage or destroy agriculture and fisheries.
- ✰ Disruption of habitats such as coral reefs and Alpine meadows could drive many plant and animal species to extinction.

Allergies, asthma, and infectious disease outbreaks will become more common due to increased growth of pollen-producing ragweed, higher levels of air pollution, and the spread of conditions favorable to pathogens and mosquitoes.

4) Ozone Layer Depletion

Ozone layer depletion is one of the most serious problems faced by our planet earth. It is also one of the prime reasons which are leading to global warming. Ozone is a colourless gas which is found in the stratosphere of our upper atmosphere. The layer of ozone gas is what which protects us from the harmful ultraviolet radiations of the sun. The ozone layer absorbs these harmful radiations and thus prevents these rays from entering the earth's atmosphere.

The main things that lead to destruction of the ozone gas in the ozone layer is low temperatures, increase in the level of chlorine and bromine gases in the upper stratosphere. But the one and the most important reason for ozone layer depletion is the production and emission of chlorofluorocarbons (CFCs). This is what which leads to almost 80 per cent of the total ozone layer depletion.

Methods to Control Air Pollutants

Methods of controlling gaseous air pollutants:

(i) **Combustion:** This technique is used for controlling those air pollutants that are in the form of organic gases or vapours. In this technique, the organic air pollutants are subjected to flame combustion technique (also known as catalytic combustion). In this technique, organic pollutants are converted into less harmful products and water vapour.

(ii) **Absorption:** Absorption is a process in which a substance penetrates into another substance like scrubbers. In this technique, gaseous pollutants are passed through absorbing material like scrubbers. These scrubbers contain a liquid absorbent. This liquid absorbent removes the pollutants present in gaseous effluents. Thus the air coming into scrubber is free from pollutants and it is discharged into atmosphere.

(iii) **Adsorption:** Adsorption is a process in which a substance sticks to the surface of another substance (called absorbent). In this technique, gaseous effluents are passed through porous solid absorbent kept in containers. The gaseous pollutants stick to the surface of the porous material and clean air passes through. The organic and inorganic constituents of gaseous effluents are trapped at the interface of solid adsorbent by physical adsorbent.

Prevention and Control of Industrial Pollution

Industrial pollution can be greatly reduced by: (a) use of cleaner fuels such as liquefied natural gas (LNG) in power plants, fertilizer plants, etc. which is cheaper in addition to being environmentally friendly, (b) employing environment friendly industrial processes so that emission of pollutants and hazardous waste is minimized, (c) installing devices which reduce release of pollutants. Devices like filters, electrostatic precipitators, inertial collectors, scrubbers are described below:

(i) **Filters –** Filters remove particulate matter from the gas stream. The medium of a filter may be made of fibrous materials like cloth, granular material like sand, a rigid material like screen, or any mat-like felt pad. Baghouse filtration system is the most common one and is made of cotton or synthetic fibers (for low temperatures) or glass cloth fabrics (for higher temperature up to 290°C).

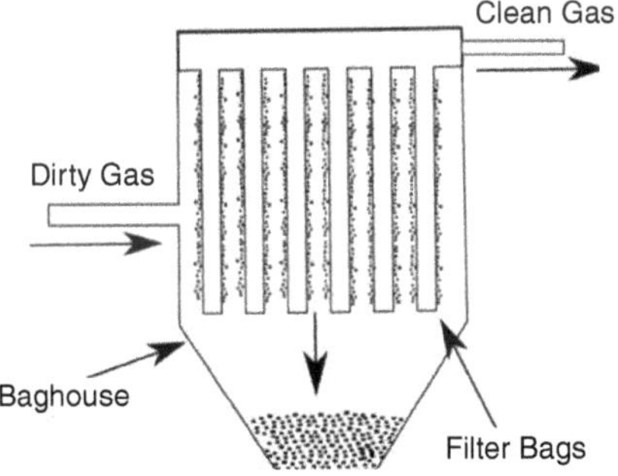

Figure 13.4. Baghouse filter

(ii) **Electrostatic precipitators (ESP)–**The emanating dust is charged with ions and the ionized particulate matter is collected on an oppositely charged surface. The particles are removed from the collection surface by occasional shaking or by rapping the surface. ESPs are used in boilers, furnaces, and many other units of thermal power plants, cement factories, steel plants, etc.

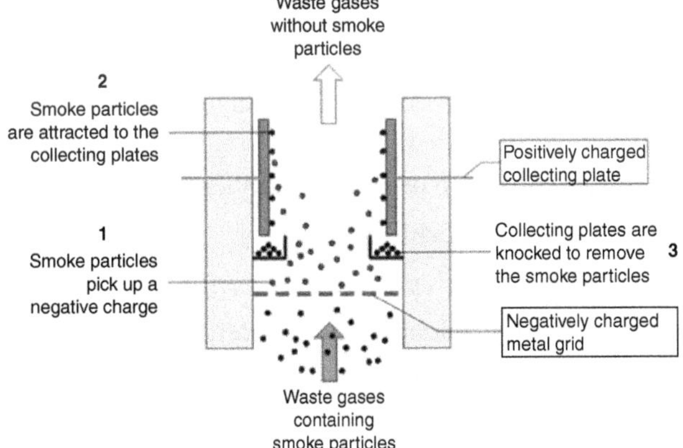

Figure 13.5. Electrostatic Precipitator

(iii) **Inertial collectors–** It works on the principle that inertia of SPM in a gas is higher than its solvent and as inertia is a function of the mass of the particulate matter this device collects heavier particles more efficiently. 'Cyclone' is a common inertial collector used in gas cleaning plants.

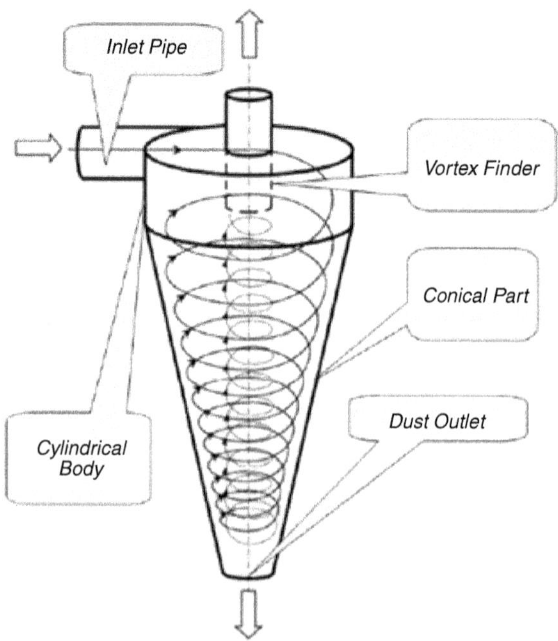

Figure. 13.6. Cyclone Dust collector

(iv) **Scrubbers–** Scrubbers are wet collectors. They remove aerosols from a stream of gas either by collecting wet particles on a surface followed by their removal, or else the particles are wetted by a scrubbing liquid. The particles get trapped as they travel from supporting gaseous medium across the interface to the liquid scrubbing medium. Gaseous pollutants can be removed by absorption in a liquid using a wet scrubber and depends on the type of the gas to be removed, e.g., for removal of sulphur dioxide alkaline solution is needed as it dissolves sulphur dioxide. Gaseous pollutants may be absorbed on an activated solid surface like silica gel, alumina, carbon, etc. Silica gel can remove water vapour. Condensation allows the recovery of many by-products in coal and petroleum processing industries from their liquid effluents.

Apart from the use of above mentioned devices, other control measures are-

- ✰ Increasing the height of chimneys.
- ✰ Closing industries which pollute the environment.
- ✰ Shifting of polluting industries away from cities, and
- ✰ Development and maintenance of green belt of adequate width.

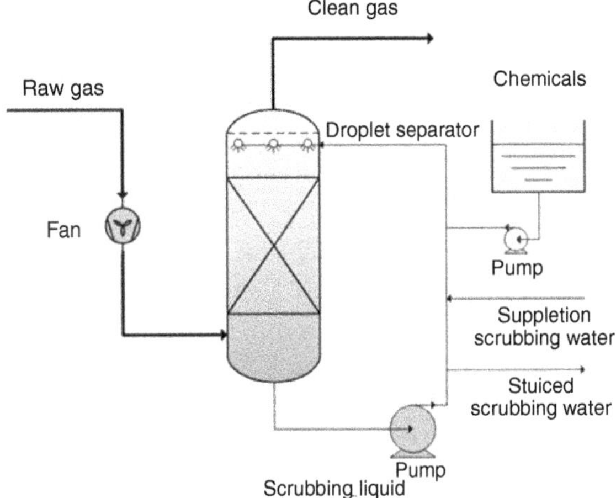

Figure. 13.7. Gas Scrubbing Instrument

Control of Vehicular Pollution

- The emission standards for automobiles have been set which if followed will reduce the pollution. Standards have been set for the durability of catalytic converters which reduce vehicular emission.
- In cities like Delhi, motor vehicles need to obtain Pollution Under Control (PUC) certificate at regular intervals. This ensures that levels of pollutants emitted from vehicle exhaust are not beyond the prescribed legal limits.
- The price of diesel is much cheaper than petrol which promotes use of diesel. To reduce emission of sulphur dioxide, sulphur content in diesel has been reduced to 0.05%.
- Earlier lead in the form of tetraethyl lead was added in the petrol to raise octane level for smooth running of engines. Addition of lead in petrol has been banned to prevent emission of lead particles with the vehicular emission.

Indoor Air Pollution

When a building is not properly ventilated, pollutants can accumulate and reach concentrations greater than those typically found outside. This problem has received media attention as "Sick Building Syndrome". Environmental tobacco smoke (ETS) is one of the main contributors to indoor pollution, as are CO, NO, and SO_2, which can be emitted from furnaces and stoves. Cleaning or remodeling a house is an activity that can contribute to elevated concentrations of harmful chemicals such as VOCs emitted from household cleaners, paint, and varnishes. Also, when bacteria die, they release endotoxins into the air, which can cause adverse health effects. So ventilation is important while cooking, cleaning, and disinfecting in a building. A geogenic source of indoor air pollution is radon.

Pollutant	Description	Natural Sources	Anthropogenic Sources	Effects
Carbon Monoxide (CO)	CO is an odorless, colorless, and poisonous gas produced by the incomplete burning of fossil fuels (gasoline, oil, natural gas).	Atmospheric oxidation of methane and other biogenic hydrocarbons	Combustion of fossil fuels	CO interferes with the blood's ability to carry oxygen, slowing reflexes and causing drowsiness. In high concentrations, CO can cause death. Headaches and stress on the heart can result from exposure to CO.
CO_2	Colorless, odorless nontoxic gas moderately soluble in water		Combustion of fossil fuels	Global warming lead to kidney desease
NO	Colorless, odorless gas; nonflammable and slightly soluble in water; toxic			NO_x can make the body vulnerable to respiratory infections, lung disease, and possibly cancer.
NO_2	Reddish-orange-brown gas with sharp, pungent odor; toxic and highly corrosive			NO_2 contributes to the brownish haze seen over congested areas and to acid rain. NO_2 easily dissolves in water and forms acids which can cause metal corrosion and fading/ deterioration of fabrics.
NH_3	Colorless gas with pungent odor	Bacterial decomposition of amino acids in organic waste	Combustion	Toxic at high concentration. In human could cause eye irritation, nose and throat as well as burning of skin.
Sulfur Dioxide (SO_2)	Colorless gas with irritating, pungent odor	Atmospheric oxidation of organic sulfides	Fuel combustion in stationary sources; industrial process emissions; metal and petroleum refining	SO_2 easily dissolves in water and forms an acid which contributes to acid rain. Lakes, forests, metals, and stone can be damaged by acid rain.
Ozone (O_3)	Colorless, toxic gas, slightly soluble in water	Natural tropospheric chemistry; transport from stratosphere to troposphere	No primary sources; formed as a secondary pollutant from atmospheric reactions involving hydrocarbons and oxides of nitrogen	Sunburn on skin throat irritation coughing and pains.

Case Studies on Air Pollution

Bhopal Gas Tragedy

On December 3, 1984, more than 40 tons of Methyl Isocyanate (MIC) gas leaked from a pesticide plant Union Carbide Company in Bhopal, MP, India, immediately killing at least 3,800 people and causing significant morbidity and premature death for many thousands more. The chemical spill turned the UCIL factory into a gas chamber. The people were running, dying and vomiting. It is considered to be the world's worst industrial disaster. Thousands of people died of primary causes of death was choking, reflexogenic circulatory collapse pulmonary oedema. In the immidiate alternate, the plant was closed to outsiders by Indian government.

Taj Mahal Damage

On the time scale of several years the outer marble surfaces of the Taj Mahal become discolored and must be cleaned in a time-consuming process. Local air quality is responsible and suggestions have included surface reactions with gas-phase SO2, as well as aqueous phase chemistry linked with the deposition of fog droplets, and water condensation, as well as dust deposition. Many measures have been undertaken to avoid the impact of local air pollution, including restricting traffic within 1 km of the grounds and limiting the emissions of industrial pollution in the city of Agra and the Mathura refinery is being shifted.

Questions

Short Answer Type Questions

1. Define pollutant.
2. Classify pollutants.
3. Enlist natural sources of air pollution.
4. Write the sources of suspended particulate matter in atmosphere.
5. What are the local effects of air pollution on environment?

Essay Type Questions

1. Define pollution. Explain various types of pollutions.
2. Define air pollution. What are the main sources of air pollution?
3. What are the impacts of air pollution on human health?
4. How air pollution affects vegetation?
5. Discuss various methods of controlling air pollution.

14

Soil Pollution

The top fertile layer of the earth's crust, which can support the vegetation growth is known as soil.

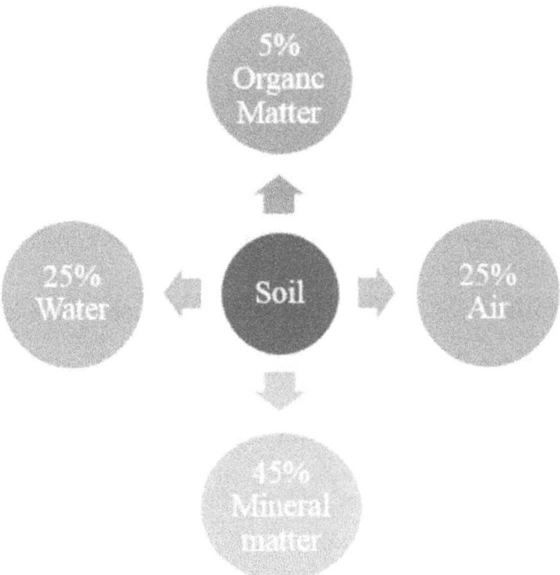

Figure 14.1. Composition of Soil

Formation of the Soil

First step: Physical weathering of bedrock, that is, the fragmentation of rocks due to temperature, wind and water.

Second step: Chemical weathering, in which the rocks are dissolved in water-soluble minerals and nutrients. Clay, as the main soil composition is formed.

Third step: Biological phase of soil formation in which lower and higher organisms settle in the physical and chemical detritus and after decomposing create the mold.

Factors Affecting the Formation of Soil

All the factors together, form the soil and they cannot replace each other, but some may only temporary and locally dominate.

1. **Soil forming rocks:** The raw material for soil formation is provided by the rocks.
2. **Climatic factors:** Climatic factors include:
   - **Temperature:** indicates how much energy comes to the surface and to what extent and for how long physical and chemical processes assist in the formation of the soil, and determines to what plants can live in the soil.
   - **Rainfall:** determines the amount and form of water coming to the surface and thus indirectly the weathering conditions, as well as affects the plants on the surface.
   - **Wind:** indirect effect is increasing of evaporation and transpiration, while the direct effect is deflation, namely soil degradation caused by the wind.
3. **Topography:** Determine the surface and subsurface water movements, material and energy flow processes. They play a role in modifying the impact of climatic and geological factors, e.g., with increasing altitude temperature decreases, while rainfall increases; it affects the erosion, i.e. soil degradation caused by water.
4. **Biological factors:** It involves activities of surface and soil organisms. This is called biological weathering.
5. **Age of soils:** Physical, chemical and biological processes take time.

Human Activities Affecting Soil Formation

Human activities alter the impact of the natural factors prevailing in soil formation. Human activities, such as a modifying factor also a contribute to soil formation. Correctly applied interventions:increases fertility, faulty management may reduce, or even ruin the soil (e.g., environmental pollution, soil acidification, etc.)

Soil Properties

Physical properties

Structure of the soil: Greater or lesser elements of the soil occur partially stuck together partly with humus or other binders stuck.

Most frequent soil structure forms: sandy; dusty; crumbly (like a porous sponge form); coarsely crumble or rough (dense, saturated with calcium, but not bound with humus); laminar (parallel layer structure); polyhedral (rich in mud, moist, during airless conditions); columns (typical for neutral saline); walnut (rounded edges and corners nuggets on the effect of tree roots).

Chemical Properties

1. **Colour:** Fe^{3+} content: gives a tan colour; $iron^{2+}$ content: gives a bluish, greenish-grayish colour; Mn^{3+} content: gives a purplish colour; $CaCO_3$: gives a light colour; humus: dark stains in the soil;
2. **PH**: expresses H^+ concentration; neutral (7): (6.8-7.2) alkaline, below this acidic.
3. **Soil Nutrients**: Soil nutrients includes all the micro- and mocro nutrients required by the plants present in the soil.

Soil Nutrients

Biological Properties

1. Micro-organisms
2. Insects
3. Animals

Soil Pollution

"Soil pollution" refers to the presence of a chemical or substance out of place and/or present at a higher than normal concentration that has adverse effects on any non-targeted organism (FAO and ITPS, 2015). Although the majority of pollutants have anthropogenic origins, some contaminants can occur naturally in soils as components of minerals and can be toxic at high concentrations. Soil pollution often cannot be directly assessed or visually perceived, making it a hidden danger. The diversity of contaminants is constantly evolving due to agrochemical and industrial developments. This diversity, and the transformation of organic compounds in soils by biological activity into diverse metabolites, makes soil surveys to identify the contaminants both difficult and expensive. The effects of soil contamination also depend on soil properties since these control the mobility, bioavailability, and residence time of contaminants.

Types of Soil Pollution

Soil pollution may be any chemicals or contaminants that harm living organisms. Pollutants decrease soil quality and also disturb the soil's natural composition and also lead to erosion of soil. Types of soil pollution can be distinguished by the source of the contaminant and its effects of the ecosystem. Types of soil pollution may be agricultural pollution, industrial pollution and urban pollution.

Agricultural Pollution

- ☆ Agricultural processes contribute to soil pollution.
- ☆ Fertilizers increase crop yield and also cause pollution that impacts soil quality.
- ☆ Pesticides also harm plants and animals by contaminating the soil.
- ☆ These chemicals get deep inside the soil and poison the groundwater system.
- ☆ Runoff of these chemicals by rain and irrigation also contaminate the local water system and is deposited at other locations.

Industrial Pollution

- About 90% of soil pollution is caused by industrial waste products.
- Improper disposal of waste contaminates the soil with harmful chemicals.
- These pollutants affect plant and animal species and local water supplies and drinking water.
- Toxic fumes from the regulated landfills contain chemicals that can fall back to the earth in the form of acid rain and can damage the soil profile.

Urban Pollution

- Human activities can lead to soil pollution directly and indirectly.
- Improper drainage and increase run-off contaminates the nearby land areas or streams.
- Improper disposal of trash breaks down into the soil and it deposits in a number of chemical and pollutants into the soil. These may again seep into groundwater or wash away in local water system.
- Excess waste deposition increases the presence of bacteria in the soil.
- Decomposition by bacteria generates methane gas contributing to global warming and poor air quality. It also creates foul odors and can impact quality of life.

Sources of Soil Pollution

Soil pollution, as has been said, can result from both intended and unintended activities. These activities can include the direct deposition of contaminants into the soil as well as complex environmental processes that can lead to indirect soil contamination through water or atmospheric deposition.

1) Point Sources

Soil pollution can be caused by a specific event or a series of events within a particular area in which contaminants are released to the soil, and the source and identity of the pollution is easily identified. This type of pollution is known as point-source pollution. Anthropogenic activities represent the main sources of point-source pollution. Examples include former factory sites, inadequate waste and wastewater disposal, uncontrolled landfills, excessive application of agrochemicals, spills of many types, and many others. Activities such as mining and smelting that are carried out using poor environmental standards are also sources of contamination with heavy metals in many regions of the world. Other examples of point-source pollution are aromatic hydrocarbons and toxic metals, which are related to oil products. Point-source pollution is very common in urban areas. Soils near roads have high levels of heavy metals, polycyclic aromatic hydrocarbons, and other pollutants.

2) Diffuse Sources

Diffuse pollution is pollution that is spread over very wide areas, accumulates in soil, and does not have a single or easily identified source. Diffuse pollution occurs where emission, transformation and dilution of contaminants in other media have occurred prior to their transfer to soil. Diffuse pollution involves the transport of pollutants *via* air-soil-water systems. For that reason, diffuse pollution is difficult to analyze, and it can be challenging to track and to delimit its spatial extent. Examples of diffuse pollution are numerous and can include sources from nuclear power and weapons activities; uncontrolled waste disposal and contaminated effluents released in and near catchments; land application of sewage sludge; the agricultural use of pesticides and fertilizers which also add heavy metals, persistent organic pollutants, excess nutrients and agrochemicals that are transported downstream by surface runoff; flood events; atmospheric transport and deposition; and/or soil erosion. Diffuse pollution has a significant impact on the environment and human health, although its severity and extent are generally unknown.

Causes of Soil Pollution

Soil Pollution is a result of many activities by mankind which contaminate the soil. Soil pollution is often associated with indiscriminate use of farming chemicals, such as pesticides, fertilizers, etc. Pesticides applied to plants can also leak into the ground, leaving long-lasting effects. Some of the harmful chemicals found in the fertilizers (e.g., cadmium) may accumulate above their toxic levels, ironically leading to the poisoning of crops. Heavy metals can enter the soil through the use of polluted water in watering crops, or through the use of mineral fertilizers.

Faulty landfills, bursting of underground bins and seepage from faulty sewage systems could cause the leakage of toxins into the surrounding soil. Acid rains caused by industrial fumes mixing in rain falls on the land, and could dissolve away some of the important nutrients found in soil, such as change the structure of the soil. Industrial wastes are one of the biggest soil-pollution factors. Iron, steel, power and chemical manufacturing plants which irresponsibly use the Earth as a dumping ground often leave behind lasting effects for years to come.

Fuel leakages from automobiles, which get washed by rain, can seep into the nearby soil, polluting it. Deforestation is a major cause for soil erosion, where soil particles are dislodged and carried away by water or wind. As a result, the soil loses it structure as well as important nutrients found in the soil. Some of the causes of soil pollution can be as follows:

- Industrial effluents like harmful gases and chemicals.
- Use of chemicals in agriculture like pesticides, fertilizers and insecticides.
- Improper or ineffective soil management system.
- Unfavorable irrigation practices.
- Improper management and maintenance of septic system.
- Sanitary waste leakage.
- Toxic fumes from industries get mixed with rains causing acid rains.
- Leakages of fuel from automobiles are washed off due to rains and are deposited in the nearby soil.
- Unhealthy waste management techniques release sewage into dumping grounds and nearby water bodies.

- ✰ Use of pesticides in agriculture retains chemicals in the environment for a long time. These chemicals also effect beneficial organisms like earthworm in the soil and lead to poor soil quality.
- ✰ Absence of proper garbage disposal system leads to scattered garbage in the soil. These contaminants can block passage of water into the soil and affects its water holding capacity.
- ✰ Unscientific disposal of nuclear waste contaminate soil and can cause mutations.
- ✰ Night soil contamination due to improper sanitary system in villages can cause harmful diseases.

Sources of Soil Pollutants

1) Natural or Geogenic Sources

Natural concentrations in the soils of a region will be strongly related to the pedo-geochemical fraction and the dynamics of the environment that led to the formation of the soil. For example, heavy metals in soils can vary over two to three orders of magnitude, considering the natural variation in the concentration of trace metals within the parent rock type. Several soil parent materials are natural sources of certain heavy metals and other elements, such as radionuclides, and these can pose a risk to the environment and human health at elevated concentrations. Arsenic (As) contamination is one of the major environmental problems around the world. Natural sources of As include volcanic releases and weathering of As-containing minerals and ores, but also naturally occurring mineralized zones of arsenopyrite (gossans), formed by the weathering of sulphide-bearing rock. Many of these minerals present a high spatial variability and many of them can be found in higher concentrations in deeper layers. However, As is slightly bio accessible when coming from natural sources. Soils and rocks are also natural sources of the radioactive gas Radon (Rn).

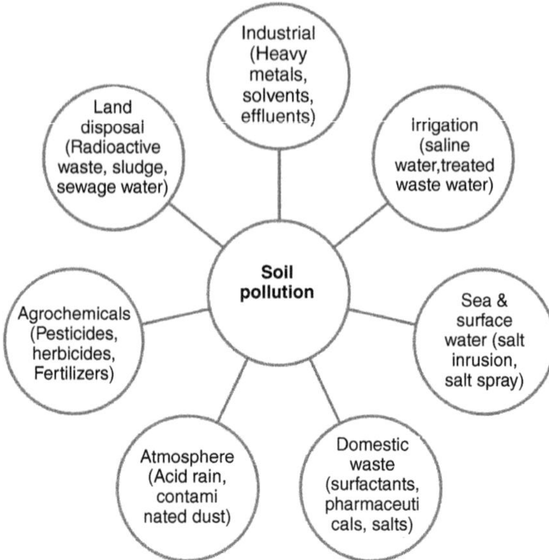

Figure 14.2. Sources of Soil Pollution

2) Anthropogenic Sources

The main anthropogenic sources of soil pollution are the chemicals used in or produced as by-products of industrial activities, domestic and municipal wastes, including wastewater, agrochemicals, and petrol-derived products. These chemicals are released to the environment accidentally, for example from oil spills or leaching from landfills, or intentionally, as in the case with the use of fertilizers and pesticides, irrigation with untreated wastewater, or land application of sewage sludge.

A) Industrial Activities

The range of chemicals used in industrial activities is vast, as is their impact on the environment. Industrial activities release pollutants to the atmosphere, water and soil. Gaseous pollutants and radio-nuclides are released to the atmosphere and can enter the soil directly through acid rain or atmospheric deposition; former industrial land can be polluted by incorrect chemical storage or direct discharge of waste into the soil; water and other fluids used for cooling in thermal power plants and many other industrial processes are discharged back to rivers, lakes and oceans, causing thermal pollution and dragging heavy metals and chlorine that affect aquatic life and other water bodies. Heavy metals from anthropogenic activities are also frequent in industrial sites and can arise from dusts and spillages of raw materials, wastes, final product, fuel ash, and fires, another major threat to global soils, affects many soils which are close to certain industrial activities, mainly those associated with chlor-alkali, textiles, glass, rubber production, animal hide processing and leather tanning, metal processing, pharmaceuticals, oil and gas drilling, pigment manufacture, ceramic manufacture, and soap and detergent production.

B) Mining

Mining has had a major impact on soil, water and biota since ancient times. Many documented examples can be found of heavily contaminated soils associated with mining activities around the world. Metal smelting to separate minerals has introduced many pollutants into the soil. Mining and smelting facilities release huge quantities of heavy metals and other toxic elements to the environment; these persist for long periods, long after the end of these activities. Toxic mining wastes are stocked up in tailings, mainly formed by fine particles that can have different concentrations of heavy metals. These polluted particles can be dispersed by wind and water erosion, sometimes reaching agricultural soils.

C) Urban and Transport Infrastructures

The widespread development of infrastructure such as housing, roads and railways has considerably contributed to environmental degradation. Their more evident negative effects on soil are soil sealing and land consumption. Apart from these known soil threats, another major impact of infrastructure activities is the entry into the soil system of different pollutants. Despite its being a major threat, soil pollution from infrastructure activities has received very minor consideration in terms of planning and impact assessment. Activities linked to transportation in and around urban centers constitute one of the main sources of soil pollution, not only because of the emissions from internal combustion engines that reach soils at more than a 100 m distance by atmospheric deposition and petrol spills, but also from the activities and the changes that result from them as a whole. Splashes generated by traffic during rainfall events and runoff, which may be significant if the drainage system is not well maintained, may translocate particles rich in heavy metals from the corrosion of metal vehicle parts, tires and pavement abrasion and other pollutants such as polycyclic aromatic hydrocarbons, rubber and plastic derived compounds. The soil contamination that resulted from this was concentrated around roads and is especially high in core urban areas.

D) Waste and Sewage Generation and Disposal

As the global population increases, so does the generation of waste. In developing and least developed countries, high rates of population growth and increasing waste and sludge production, combined with lack of municipal services that deal with waste management, create a dangerous situation. Municipal waste disposal in landfills and incineration are the two most common ways to manage waste. In both cases, many pollutants, such as heavy metals, polyaromatic hydrocarbons, pharmaceutical compounds, personal care products and their derivative products accumulate in the soil, either directly from landfill leachates that may be polluting soil and underground water, or by ash fallout from incinerating plants.

E) Agricultural and Livestock Activities

The different agricultural sources of soil pollutants include agrochemical sources, such as fertilizers and animal manure, and pesticides. Trace metals from these agrochemicals, such as, Cu, Cd, Pb and Hg, are also considered soil pollutants as they can impair plant metabolism and decrease crop productivity. Water sources for irrigation can also cause soil pollution if they consist of waste water and urban sewage. Excess N and heavy metals are not only a source of soil pollution, but also a threat to food security, water quality and human health, when they enter the food chain.

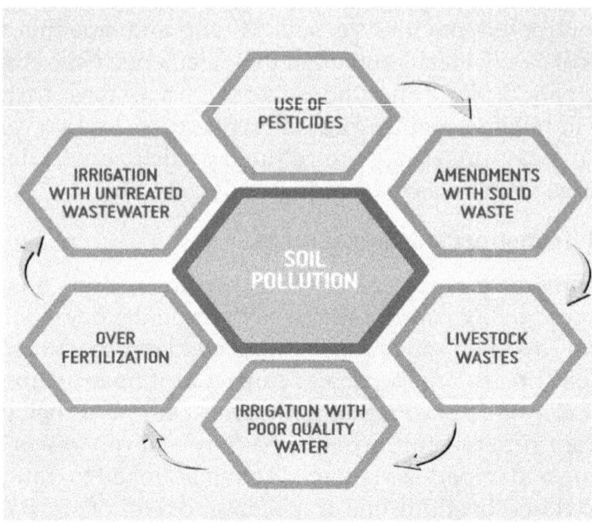

Figure 14.3. Agricultural sources of Soil Pollution

Main Pollutants in Soil

The release of pollutants to the environment, as has been mentioned, usually originates from anthropogenic processes. Even if some elements and compounds occur naturally in soils, human interventions are the main drivers of soil pollution. The following sections discuss only a small subset of the most common pollutants affecting agricultural areas, and the properties that make these compounds pollutants. Pollutants have been divided by their chemical characteristics, but some of the categories presented here overlap.

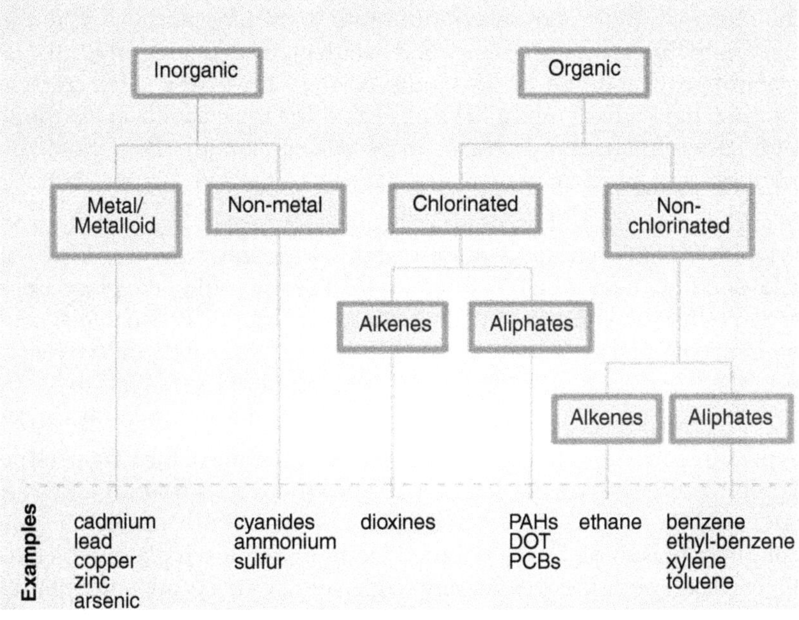

Figure 14.4. Various types of soil pollutants.

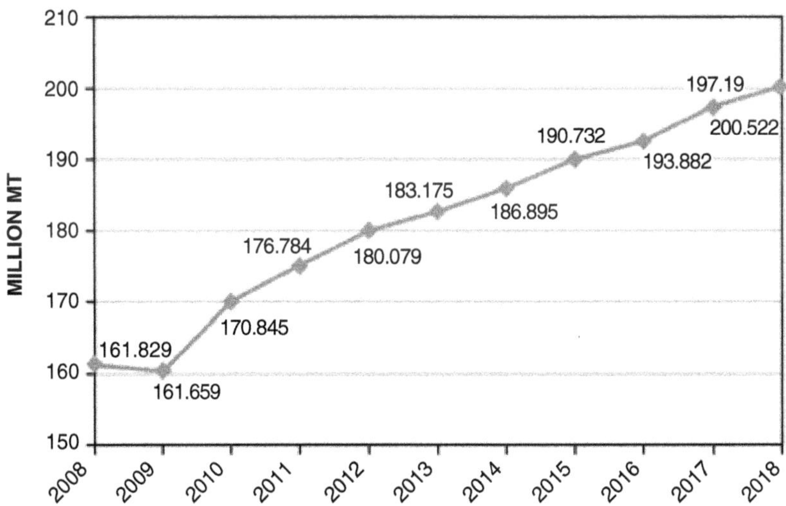

Figure 14.5. Global Consumption of synthetic Fertilizers (Source FAO)

Effects of Soil Pollution

The main reason for soil contamination is due to the presence of anthropogenic activities. These waste products are made of chemicals that are not originally found in nature and hence lead to soil pollution. Soil pollution is typically caused by industrial activity, chemicals used in agriculture and improper disposal of waste. Soil contamination leads to health risks due to direct and indirect contact with contaminated soil. Soil pollution causes huge disturbances in the ecological balance and the health of the organisms is under risk.

The effects of pollution on soil are quite disturbing and can result in huge disturbances in the ecological balance and health of living beings on earth. Normally crops cannot grow and flourish in a polluted soil. However if some crops manage to grow, then these crops might have absorbed the toxic chemicals in the soil and might cause serious health problems in people consuming them. Sometimes the soil pollution is in the form of increased salinity of the soil.

In such a case, the soil becomes unhealthy for vegetation, and often becomes useless and barren. When soil pollution modifies the soil structure, deaths of many beneficial soil organisms (e.g., earthworms) in the soil could take place. Other than further reducing the ability of the soil to support life, this occurrence could also have an effect on the larger predators (e.g., birds) and force them to move to other places, in the search of food. People living near polluted land tend to have higher incidences of migraines, nausea, fatigue, skin disorders and even miscarriages.

Depending on the pollutants present in the soil, some of the longer-term effects of soil pollution include cancer, leukemia, reproductive disorders, kidney and liver damage, and central nervous system failure. These health problems could be a result of direct poisoning by the polluted land (e.g., children playing on land filled with toxic waste) or indirect poisoning (e.g., eating crops grown on polluted land, drinking water polluted by the leaching of chemicals from the polluted land to the water supply, etc.

Long term effects of Soil pollution

The long-term effects of soil pollution are many and can be difficult to deal with, depending on the nature of the contamination.

Soil is a sort of ecosystem unto itself, and it is relatively sensitive to foreign matter being applied to it. That's good for us in the case of wanting to add soil amendments, fertilizer and compost to make the soil healthier, but not so good when it comes to soil pollution. There are many different ways that soil can become polluted, such as:

- Seepage from a landfill
- Discharge of industrial waste into the soil.
- Percolation of contaminated water into the soil.
- Rupture of underground storage tanks.
- Excess application of pesticides, herbicides or fertilizer.
- Solid waste seepage.

The most common chemicals involved in causing soil pollution are:

- ✰ Petroleum hydrocarbons
- ✰ Heavy metals
- ✰ Pesticides
- ✰ Solvents

Soil pollution happens when these chemicals adhere to the soil, either from being directly spilled onto the soil or through contact with soil that has already been contaminated. As the world becomes more industrialized, the long-term effects of soil pollution are becoming more of a problem all over the world. Even when soil is not being used for food, the matter of its contamination can be a health concern. This is especially so when that soil is found in parks, neighborhoods or other places where people spend time.

Health effects will be different depending on what kind of pollutant is in the soil. It can range from developmental problems, such as in children exposed to lead, to cancer from chromium and some chemicals found in fertilizer, whether those chemicals are still used or have been banned but are still found in the soil. Some soil contaminants increase the risk of leukemia, while others can lead to kidney damage, liver problems and changes in the central nervous system. Those are just the long-term effects of soil pollution.

In the short-term, exposure to chemicals in the soil can lead to headaches, nausea, fatigue and skin rashes at the site of exposure. When it comes to the environment itself, the toll of contaminated soil is even direr. Soil that has been contaminated should no longer be used to grow food, because the chemicals can leech into the food and harm people who eat it. If contaminated soil is used to grow food, the land will usually produce lower yields than it would if it were not contaminated.

This, in turn, can cause even more harm because a lack of plants on the soil will cause more erosion, spreading the contaminants onto land that might not have been tainted before. In addition, the pollutants will change the makeup of the soil and the types of microorganisms that will live in it. If certain organisms die off in the area, the larger predator animals will also have to move away or die because they've lost their food supply. Thus it's possible for soil pollution to change whole ecosystems.

1) Effects on Soil Micro-organism

The effects of pesticides on soil micro organisms can cause a ripple effect that can last for years. Micro-organisms are essential to healthy soil. Without them, your plants will not reach their true potential. Micro-organisms are organisms that are too small to be seen with the human eye. They live on the top-most layer of soil. There are many micro-organisms which live in the soil including:

- ✰ Bacteria
- ✰ Fungi
- ✰ Algae
- ✰ Protozoa

Micro-organisms are responsible for the decomposition and recycling of organic materials in the soil. They aid in the plant's absorption of essential nutrients. An example of this is the nitrogen fixing bacteria, *Bradyrhizobium*, which lives in a nodule on the soybean plant. It provides nitrogen to the plant and boosts growth. Biopesticides are micro-organisms that can help a plant defend it against pests. These micro-organisms include antimicrobial metabolites, antibiotics and extracellular enzymes. The potential of these biopesticides has not been fully examined by scientists. It is hopeful that science will be able to re-produce the effects of the biopesticides.

Alternatives to Harmful Chemical Pesticides

For the average gardener, the use of organic pesticides can keep a healthy balance in the soil. Many organic pesticides are made of minerals or other plant materials that will keep pests at bay and break down quickly in the soil. Examples of some common organic pesticides include the following:

- **Cayenne pepper spray-**Can be sprayed on the leaves of plants to deter harmful insects.
- **Soap spray-**Also sprayed on plants to get rid of aphids.
- **Tobacco powder-**A spray can be made from the finely ground tobacco leaves and water. It is used to kill sucking insects on plants such as aphids, thrips and spider mites.
- **Pyrethrin-**Made from the chrysanthemum plant. This organic pesticide is used to knock out flying insects and ground pests such as grubs.
- **Neem-**Derived from the neem tree. Used to control Gypsy moths, leaf miners, mealy bugs, whiteflies and caterpillars.
- **Sabadilla-**Derived from the sabadilla lily. Used to control caterpillars, leaf hoppers, stink bugs and squash bugs.

2) Some other Effects of Soil Pollution are

- Disturbance and imbalance in the of flora and fauna inhabiting in the soil.
- Contaminated soil decreases soil fertility and hence there is decrease in the soil yield.
- Reduced soil fertility hence decrease in crop yield.
- Loss of natural nutrients in soil.
- Reduced nitrogen fixation.
- Increased soil erosion.
- Increase in soil salinity, makes it unfit for cultivation.
- Creation of toxic dust.
- Crops grown on polluted soil cause health problems on consumption.

- ☆ Foul odor due to chemicals and gases can lead to problems like headaches, nausea, etc.
- ☆ Pollutants in soil cause alteration in soil structure, causing death of many soil organisms. This can affect the food chain.

3) Effects on Humans

Soil pollution has major consequences on human health. Consumption of crops and plants grown on polluted soil cause health hazards.

- ☆ Long term exposure to polluted soil affects the genetic make-up of the body and may cause congenital illnesses and chronic health diseases.
- ☆ Chronic exposure to heavy metals, petroleum, solvents and agricultural chemicals can be carcinogenic.
- ☆ The foul odor causes inconvenience to people.

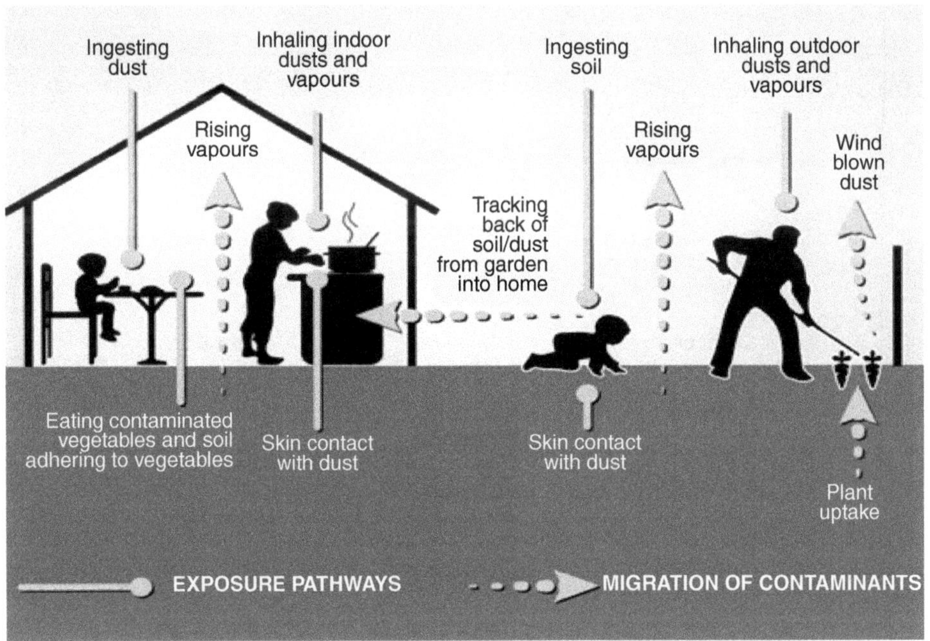

Figure 14.6. Health effects of Soil Pollution

The route of human exposure to a soil contaminant will vary depending on the contaminant itself and on the conditions and activities at a particular site. Generally, people can be exposed to contaminants present in soil through ingestion or through the consumption of plants or animals that have accumulated large amounts of soil pollutants; through dermal exposure, from using spaces such as parks and gardens; or by inhaling soil contaminants that have been vaporized. Humans may also be affected as a result of secondary contamination of water supplies and from deposition of air contaminants.

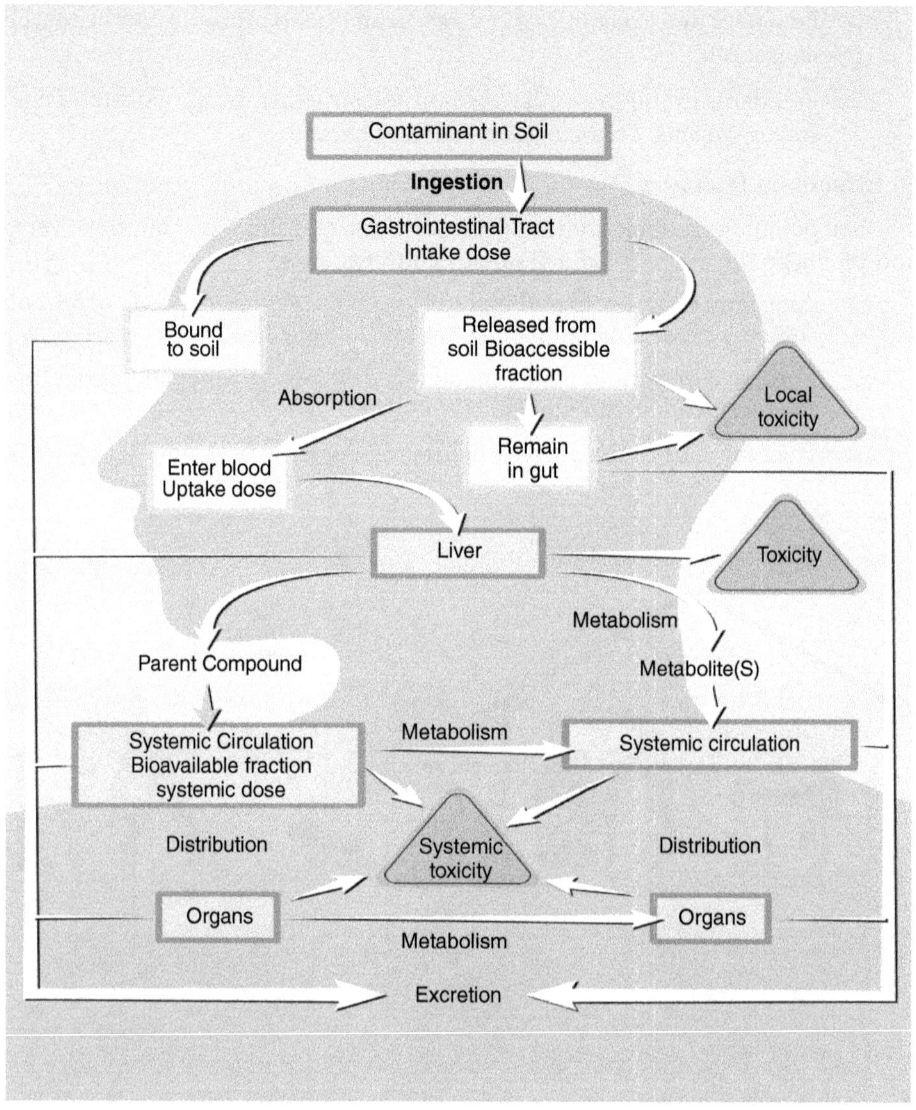

Figure 14.7. The route of human exposure to a soil contaminant

4) Effect on Soil Structure

- Soil pollution can lead to death of many soil organisms like the earthworms which can lead to alteration in the soil structure.
- This can force other predators to move to other places in search of food.

There are some ways to get soil back to its pristine condition or to remove the spoiled soil so the land can be used for agriculture again. Tainted soil can be transported to a site where humans won't be exposed to the chemicals, or the soil can be aerated to remove some of the chemicals (which can add the problem of air pollution chemicals can be released into the air). Other options include what's known as bioremediation, where microorganisms are used to consume the pollution compounds as well as electromechanical systems for extracting chemicals, an containment of chemicals by paving over the tainted area. None of these are an ideal solution. Preventing contamination in the first place is the best way to go. It won't eliminate all potential pollution problems, but choosing to farm organically is a good way to protect the soil (and yourself) from chemicals found in pesticides and other common garden chemicals.

Management and Remediation of Polluted Soils

The first step in the assessment and management of polluted soils is the identification of the problem. In general, when an area is affected by an accident such as an oil spill, a nuclear accident, or the rupture of a dam tailing, measures to control the extent and prevent further occurrences generally start immediately. However, in legacy polluted soils or where diffuse pollution could be an issue, there are often no established protocols to be followed. In some countries or regions in the world, there are national, regional or local agencies who are responsible for initiating a preliminary investigation to determine whether or not pollution is present and whether further action is needed, while there are many others where no regulation or protocols have been defined.

Control of Soil Pollution

A number of ways have been suggested to curb the pollution rate. Attempts to clean up the environment require plenty of time and resources. Some of the steps to reduce soil pollution are:

- Ban on use of plastic bags below 20 microns thickness, which are the major cause of pollution.
- Recycling of plastic wastes.
- Encouraging plantation programmes.
- Encouraging social and agroforestry programmes.
- Recycling paper, plastics and other materials.
- Reusing materials.
- Avoiding deforestation and promoting forestation.
- Suitable and safe disposal of including nuclear wastes.
- Chemical fertilizers and pesticides should be replaced by organic fertilizers and pesticides.
- Undertaking many pollution awareness programs.

Prevention of Soil Pollution

Toxic chemical compounds like salts, radioactive agents, toxins and other waste contribute to soil pollution. These have adverse effect on plant and animal health. Soil contains both organic as well as inorganic material. Other common methods of preventing soil pollution include reforestation and recycling of waste materials. Deforestation or the cutting down of trees often leads to erosion of the soil, which leads to soil pollution due to the loss of fertility of the soil. Thus, reforestation is an effective method of preventing soil pollution. In addition, reducing the volume of refuse or waste in landfills by recycling materials such as plastics, papers and various other materials is another effective and common method of preventing the phenomenon of soil pollution.

Some preventice measures to reduce soil pollition are as follows:

- ✰ Strong regulatory programs to minimize soil contamination need to be introduced.
- ✰ Reuse and recycle of unwanted items or even better, reduce consumption and reduce your trash. The less rubbish we create the less chance the waste will end up in our soil.
- ✰ There is a need to educate the public about the harms done when they litter.
- ✰ For gardens, make use of organic fertilizers and organic pesticides, because they are usually made of natural substances, are bio-degradable and do little harm to the natural balance in the soil. The common methods to control weed growth is covering the soil with layers of newspapers.

Questions

Short Answer Type Questions

1. Define soil pollution.
2. Name the physical properties of soil.
3. What are the chemical and biological properties of soil?
4. What is weathering?
5. Define soil horizons.

Essay Type Questions

1. What are the causes of soil pollution?
2. Describe the various effects of soil pollution.
3. What are the measures to be taken for the control of soil pollution?
4. What are agrochemicals and how they effect environment?

15

Soil Erosion

Soil erosion is a naturally occurring process that affects all landforms. In agriculture, soil erosion refers to the wearing away of a field's topsoil by the natural physical forces of water and wind or through forces associated with farming activities such as tillage.

Erosion, whether it is by water, wind or tillage, involves three distinct actions — soil detachment, movement and deposition. Topsoil, which is high in organic matter, fertility and soil life, is relocated elsewhere "on-site" where it builds up over time or is carried "off-site" where it fills in drainage channels. Soil erosion reduces cropland productivity and contributes to the pollution of adjacent watercourses, wetlands and lakes.

Soil erosion can be a slow process that continues relatively unnoticed or can occur at an alarming rate, causing serious loss of topsoil. Soil compaction, low organic matter, loss of soil structure, poor internal drainage, salinization and soil acidity problems are other serious soil degradation conditions that can accelerate the soil erosion process.

This looks at the causes and effects of water, wind and tillage erosion on agricultural land.

I. Water Erosion

The widespread occurrence of water erosion combined with the severity of on-site and off-site impacts have made water erosion the focus of soil conservation efforts.

The rate and magnitude of soil erosion by water is controlled by the following factors:

1) Rainfall and Runoff

The greater the intensity and duration of a rainstorm, higher is the erosion potential. The impact of raindrops on the soil surface can break down soil aggregates and disperse the aggregate material. Lighter aggregate materials such as very fine sand, silt, clay and organic matter are easily removed by the raindrop splash and runoff water; greater raindrop energy or runoff amounts are required to move larger sand and gravel particles.

Soil movement by rainfall (raindrop splash) is usually greatest and most noticeable during short duration, high-intensity thunderstorms. Although the erosion caused by long-lasting and less-intense storms is not usually as spectacular or noticeable as that produced during thunderstorms, the amount of soil loss can be significant, especially when compounded over time.

Surface water runoff occurs whenever there is excess water on a slope that cannot be absorbed into the soil or is trapped on the surface. Reduced infiltration due to soil compaction, crusting or freezing increases the runoff. Runoff from agricultural land is greatest during spring months when the soils are typically saturated, snow is melting and vegetative cover is minimal.

2) Soil Erodibility

Soil erodibility is an estimate of the inability of soils to resist erosion, based on the physical characteristics of each soil. Texture is the principal characteristic factor affecting erodibility, but structure, organic matter and permeability also contribute. Generally, soils with faster infiltration rates, higher levels of organic matter and improved soil structure have a greater resistance to erosion. Sand, sandy loam and loam-textured soils tend to be less erodible than silt, very fine sand and certain clay-textured soils.

Tillage and cropping practices that reduce soil organic matter levels, cause poor soil structure, or result in soil compaction, contribute to increases in soil erodibility. As an example, compacted subsurface soil layers can decrease infiltration and increase runoff. The formation of a soil crust, which tends to "seal" the surface, also decreases infiltration. On some sites, a soil crust might decrease the amount of soil loss from raindrop impact and splash; however, a corresponding increase in the amount of runoff water can contribute to more serious erosion problems.

Past erosion also has an effect on a soil's erodibility. Many exposed subsurface soils on eroded sites tend to be more erodible than the original soils were because of their poorer structure and lower organic matter. The lower nutrient levels often associated with subsoils contribute to lower crop yields and generally poorer crop cover, which in turn provides less crop protection for the soil.

3) Slope Gradient and Length

The steeper and longer the slope of a field, the higher the risk for erosion. Soil erosion by water increases as the slope length increases due to the greater accumulation of runoff. Consolidation of small fields into larger ones often results in longer slope lengths with increased erosion potential, due to increased velocity of water, which permits a greater degree of scouring (carrying capacity for sediment).

4) Cropping and Vegetation

The potential for soil erosion increases if the soil has no or very little vegetative cover of plants and/ or crop residues. Plant and residue cover protects the soil from raindrop impact and splash, tends to slow down the movement of runoff water and allows excess surface water to infiltrate.

The erosion-reducing effectiveness of plant and/or crop residues depends on the type, extent and quantity of cover. Vegetation and residue combinations that completely cover the soil and intercept all falling raindrops at and close to the surface are the most efficient in controlling soil erosion (e.g., forests, permanent grasses). Partially incorporated residues and residual roots are also important as these provide channels that allow surface water to move into the soil.

The effectiveness of any protective cover also depends on how much protection is available at various periods during the year, relative to the amount of erosive rainfall that falls during these periods. Crops that provide a full protective cover for a major portion of the year (e.g., alfalfa or winter cover crops) can reduce erosion much more than can crops that leave the soil bare for a longer period of time (e.g., row crops), particularly during periods of highly erosive rainfall such as spring and summer. Crop management systems that favour contour farming and strip-cropping techniques can further reduce the amount of erosion. To reduce most of the erosion on annual row-crop land, leave a residue cover greater than 30% after harvest and over the winter months, or inter-seed a cover crop (e.g., red clover in wheat, oats after silage corn).

5) Tillage Practices

The potential for soil erosion by water is affected by tillage operations, depending on the depth, direction and timing of the type of tillage equipment and the number of passes. Generally, the less the disturbance of vegetation or residue cover at or near the surface, the more effective the tillage practice in reducing water erosion. Minimum till or no-till practices are effective in reducing soil erosion by water.

Tillage and other practices performed up and down field slopes creates pathways for surface water runoff and can accelerate the soil erosion process. Cross-slope cultivation and contour farming techniques discourage the concentration of surface water runoff and limit soil movement.

Forms of Water Erosion

1) Sheet Erosion

Sheet erosion is the movement of soil from raindrop splash and runoff water. It typically occurs evenly over a uniform slope and goes unnoticed until most of the productive topsoil has been lost. Deposition of the eroded soil occurs at the bottom of the slope or in low areas. Lighter-coloured soils on knolls, changes in soil horizon thickness and low crop yields on shoulder slopes and knolls are other indicators.

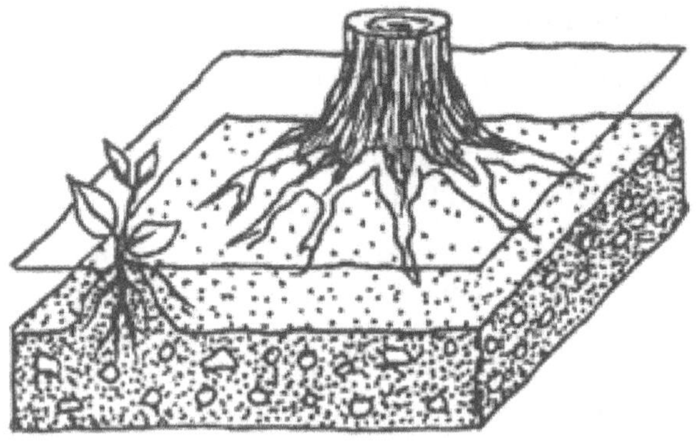

Figure 15.1. Sheet Erosion

2) Rill Erosion

Rill erosion results when surface water runoff concentrates, forming small yet well-defined channels. These distinct channels where the soil has been washed away are called rills when they are small enough to not interfere with field machinery operations. In many cases, rills are filled in each year as part of tillage operations.

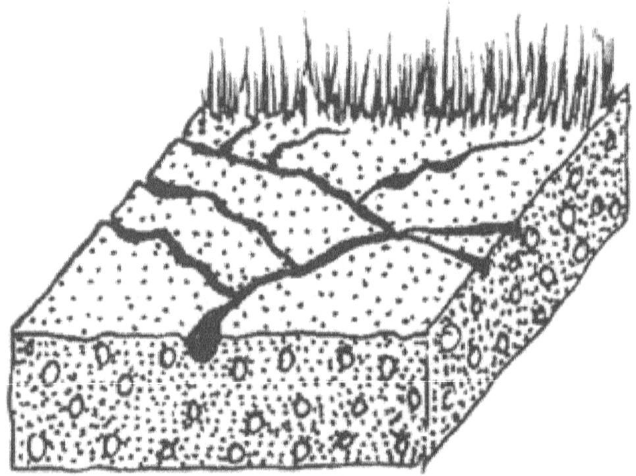

Figure 15.2. Rill Erosion

3) Gully Erosion

Gully erosion is an advanced stage of rill erosion where surface channels are eroded to the point where they become a nuisance factor in normal tillage operations. Surface water runoff, causing gully formation or the enlarging of existing gullies, is usually the result of improper outlet design for local surface and subsurface drainage systems. The soil instability of gully banks, usually associated with seepage of groundwater, leads to sloughing and slumping (caving-in) of bank slopes. Such failures usually occur during spring months when the soil water conditions are most conducive to the problem.

Gully formations are difficult to control if corrective measures are not designed and properly constructed. Control measures must consider the cause of the increased flow of water across the landscape and be capable of directing the runoff to a proper outlet. Gully erosion results in significant amounts of land being taken out of production and creates hazardous conditions for the operators of farm machinery.

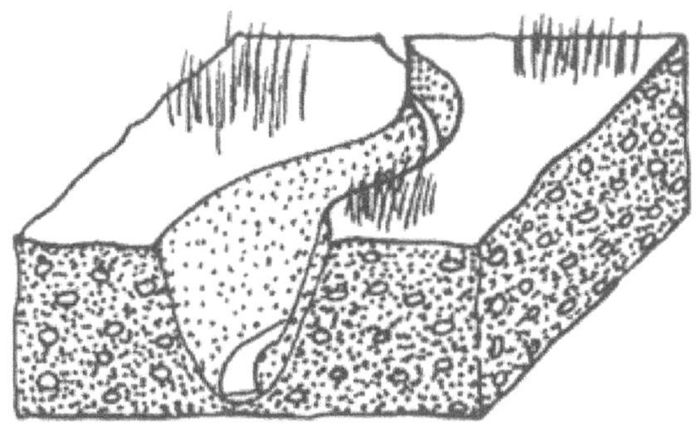

Figure 15.3. Gully Erosion

4) Bank Erosion

Natural streams and constructed drainage channels act as outlets for surface water runoff and subsurface drainage systems. Bank erosion is the progressive undercutting, scouring and slumping of these drainage ways. Poor construction practices, inadequate maintenance, uncontrolled livestock access and cropping too close can all lead to bank erosion problems.

Poorly constructed tile outlets also contribute to bank erosion. Some do not function properly because they have no rigid outlet pipe, have an inadequate splash pad or no splash pad at all, or have outlet pipes that have been damaged by erosion, machinery or bank cave-ins.

The direct damages from bank erosion include loss of productive farmland, undermining of structures such as bridges, increased need to clean out and maintain drainage channels and washing out of lanes, roads and fence rows.

Effects of Water Erosion

The implications of soil erosion by water extend beyond the removal of valuable topsoil. Crop emergence, growth and yield are directly affected by the loss of natural nutrients and applied fertilizers. Seeds and plants can be disturbed or completely removed by the erosion. Organic matter from the soil, residues and any applied manure, is relatively lightweight and can be readily transported off the field, particularly during spring thaw conditions. Pesticides may also be carried off the site with the eroded soil.

Soil quality, structure, stability and texture can be affected by the loss of soil. The breakdown of aggregates and the removal of smaller particles or entire layers of soil or organic matter can weaken the structure and even change the texture. Textural changes can, in turn, affect the water-holding capacity of the soil, making it more susceptible to extreme conditions such as drought.

The off-site impacts of soil erosion by water are not always as apparent as the on-site effects. Eroded soil, deposited down slope, inhibits or delays the emergence of seeds, buries small seedlings and necessitates replanting in the affected areas. Also, sediment can accumulate on down-slope properties and contribute to road damage.

Sediment that reaches streams or watercourses can accelerate bank erosion, obstruct stream and drainage channels, fill in reservoirs, damage fish habitat and degrade downstream water quality. Pesticides and fertilizers, frequently transported along with the eroding soil, contaminate or pollute downstream water sources, wetlands and lakes. Because of the potential seriousness of some of the off-site impacts, the control of "nonpoint" pollution from agricultural land is an important consideration.

II. Wind Erosion

Wind erosion occurs mainly is sandy and organic or muck soils. Under the right conditions it can cause major losses of soil and property.

Soil particles move in three ways, depending on soil particle size and wind strength — suspension, saltation and surface creep.

The rate and magnitude of soil erosion by wind is controlled by the following factors:

1) Soil Erodibility

Very fine soil particles are carried high into the air by the wind and transported great distances (suspension). Fine-to-medium size soil particles are lifted a short distance into the air and drop back to the soil surface, damaging crops and dislodging more soil (saltation). Larger-sized soil particles that are too large to be lifted off the ground are dislodged by the wind and roll along the soil surface (surface creep). The abrasion that results from windblown particles breaks down stable surface aggregates and further increases the soil erodibility.

2) Soil Surface Roughness

Soil surfaces that are not rough offer little resistance to the wind. However, ridges left from tillage can dry out more quickly in a wind event, resulting in more loose, dry soil available to blow. Over time, soil surfaces become filled in, and the roughness is broken down by abrasion. This results in a smoother surface susceptible to the wind. Excess tillage can contribute to soil structure breakdown and increased erosion.

3) Climate

The speed and duration of the wind have a direct relationship to the extent of soil erosion. Soil moisture levels are very low at the surface of excessively drained soils or during periods of drought, thus releasing the particles for transport by wind. This effect also occurs in freeze-drying of the soil surface during winter months. Accumulation of soil on the leeward side of barriers such as fence rows, trees or buildings, or snow cover that has a brown colour during winter are indicators of wind erosion.

4) Unsheltered Distance

A lack of windbreaks (trees, shrubs, crop residue, etc.) allows the wind to put soil particles into motion for greater distances, thus increasing abrasion and soil erosion. Knolls and hilltops are usually exposed and suffer the most.

5) Vegetation Cover

The lack of permanent vegetative cover in certain locations results in extensive wind erosion. Loose, dry, bare soil is the most susceptible; however, crops that produce low levels of residue (e.g., soybeans and many vegetable crops) may not provide enough resistance. In severe cases, even crops that produce a lot of residue may not protect the soil.

The most effective protective vegetative cover consists of a cover crop with an adequate network of living windbreaks in combination with good tillage, residue management and crop selection.

Effects of Wind Erosion

Wind erosion damages crops through sandblasting of young seedlings or transplants, burial of plants or seed, and exposure of seed. Crops are ruined, resulting in costly delays and making reseeding necessary. Plants damaged by sandblasting are vulnerable to the entry of disease with a resulting decrease in yield, loss of quality and market value. Also, wind erosion can create adverse operating conditions, preventing timely field activities.

Soil drifting is a fertility-depleting process that can lead to poor crop growth and yield reductions in areas of fields where wind erosion is a recurring problem. Continual drifting of an area gradually causes a textural change in the soil. Loss of fine sand, silt, clay and organic particles from sandy soils serves to lower the moisture-holding capacity of the soil. This increases the erodibility of the soil and compounds the problem

The removal of wind-blown soils from fence rows, constructed drainage channels and roads, and from around buildings is a costly process. Also, soil nutrients and surface-applied chemicals can be carried along with the soil particles, contributing to off-site impacts. In addition, blowing dust can affect human health and create public safety hazards.

III. Tillage Erosion

Tillage erosion is the redistribution of soil through the action of tillage and gravity. It results in the progressive down-slope movement of soil, causing severe soil loss on upper-slope positions and accumulation in lower-slope positions. This form of erosion is a major delivery mechanism for water erosion. Tillage action moves soil to convergent areas of a field where surface water runoff concentrates. Also, exposed subsoil is highly erodible to the forces of water and wind. Tillage erosion has the greatest potential for the "on-site" movement of soil and in many cases can cause more erosion than water or wind.

The rate and magnitude of soil erosion by tillage is controlled by the following factors:

1) Type of Tillage Equipment

Tillage equipment that lifts and carries will tend to move more soil. As an example, a chisel plow leaves far more crop residue on the soil surface than the conventional moldboard plow but it can move as much soil as the moldboard plow and move it to a greater distance. Using implements that do not move very much soil will help minimize the effects of tillage erosion.

2) Direction

Tillage implements like a plow or disc throw soil either up or down slope, depending on the direction of tillage. Typically, more soil is moved while tilling in the down-slope direction than while tilling in the up-slope direction.

3) Speed and Depth

The speed and depth of tillage operations will influence the amount of soil moved. Deep tillage disturbs more soil, while increased speed moves soil further.

4) Number of Passes

Reducing the number of passes of tillage equipment reduces the movement of soil. It also leaves more crop residue on the soil surface and reduces pulverization of the soil aggregates, both of which can help resist water and wind erosion.

Effects of Tillage Erosion

Tillage erosion impacts crop development and yield. Crop growth on shoulder slopes and knolls is slow and stunted due to poor soil structure and loss of organic matter and is more susceptible to stress under adverse conditions. Changes in soil structure and texture can increase the erodibility of the soil and expose the soil to further erosion by the forces of water and wind.

In extreme cases, tillage erosion includes the movement of subsurface soil. Subsoil that has been moved from upper-slope positions to lowerslope positions can bury the productive topsoil in the lower-slope areas, further impacting crop development and yield. Research related to tillage-eroded fields has shown soil loss of as much as 2 m of depth on upper-slope positions and yield declines of up to 40% in corn. Remediation for extreme cases involves the relocation of displaced soils to the upper-slope positions.

Conservation Measures

The adoption of various soil conservation measures reduces soil erosion by water, wind and tillage. Tillage and cropping practices, as well as land management practices, directly affect the overall soil erosion problem and solutions on a farm. When crop rotations or changing tillage practices are not enough to control erosion on a field, a combination of approaches or more extreme measures might be necessary. For example, contour plowing, strip cropping or terracing may be considered. In more serious cases where concentrated runoff occurs, it is necessary to include structural controls as part of the overall solution — grassed waterways, drop pipe and grade control structures, rock chutes, and water and sediment control basins.

Summary

Soil erosion remains a key challenge for agriculture. Many farmers have already made significant progress in dealing with soil erosion problems on their farms. However, because of continued advances in soil management and crop production technology that have maintained or increased yields in spite of soil erosion, others are not aware of the increasing problem on farmland. Awareness usually occurs only when property is damaged and productive areas of soil are lost.

The increase in extreme weather events predicted with climate change will magnify the existing water and wind erosion situations and create new areas of concern. Farmland must be protected as much as possible, with special attention to higher risk situations that leave the soil vulnerable to erosion.

Questions

Short Answer Type Questions

1. Define
   (a) Soil erosion
   (b) Soil erodability
   (c) Tillage erosion
2. What is saltation?

Essay Type Questions

1. Classify and explain different types of soil erosion.
2. What are the effects of soil erosion on environment?
3. Describe the conservation measures for preventing soil erosion.
4. Write down the causes of soil erosion.

16

Insecticide & Pesticide Pollution

Insecticide and Pesticide Pollution

The greatest challenge of today's agriculture is to feed the growing population and restore the natural resources. Global food production needs to be doubled by 2020 and just to maintain the present precipitate food consumption. Uncontrolled population growth in developing countries accelerated the imbalance between human needs and sustainable use of land. Though by virtue of chemical fertilizers the production and productivity of crops has increased. The increased use of pesticides has posed many environmental and health problems.

Pesticides have proved to be a boon for the farmers as well as people all around the world by increasing agricultural yield. Basically, the input of pesticides in Indian agriculture increases after the announcement of Green Revolution which in turn helps our country to fight the major problem of food crises. Although the application of pesticides serves as a boon but also had a long-term negative effect of harming the environment and human health. Currently, India is the largest producer of pesticides in Asia and ranks twelfth in the world for the use of pesticides. Although Indian average consumption of pesticide is far lower than many other developed economies, the problem of pesticide residue is very high in India.

The chemical fertilizers and pesticides used over a long period of time have adverse toxic effects on the production potential of the land and the ultimate consumers of the products. The current issue of hazard posed by pesticides to human health and the environment has raised concerns. Production of better alternative to reduce pesticide formulations is an answer to this destruction condition. If the pesticides are used in appropriate quantities and used only when required or necessary or opting for organic farming, then pesticide risks can be tackled to some extent.

Types of pesticides

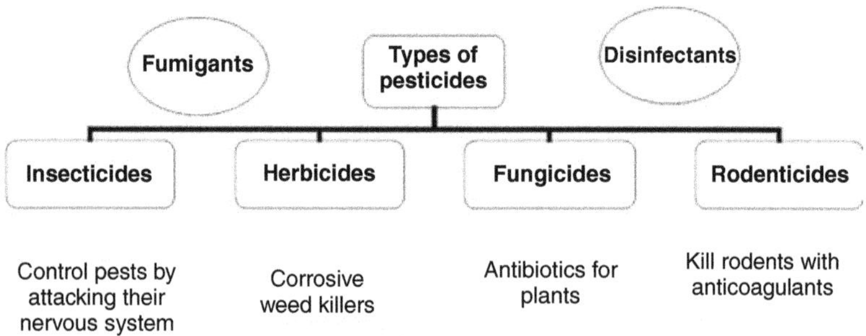

Figure 16.1. Types of Pesticides

Pesticides can be either organic or inorganic synthetic molecules. They are classified on the basis of their chemical structures, their mode of action, their way of entry into the body, and their target organisms. Their toxicological effects on pests depend on their chemical composition, which in turn affects their interaction with soil components. According to their chemical structure, pesticides can be divided into twelve distinct groups, with the main pesticides in each group listed below:

- **Organochlorine compounds:** DDT, Methoxychlor, Chlordane, Dicofol. BHC/HCH, Aldrin, Endosulfan, Heptachlor, Methoxychlor, Chlordane, Dicofol.
- **Organophosphorus compounds:** Parathion, Malathion, Monocrotophos, Chlorpyrifos, Quinalphos, Phorate, Diazinon, Fenitrothion, Acephate, Dimethoate, Fenthion, Isofenfos, Phosphamidon, Temephos, Triazophos.
- **Carbamates:** Aldicarb, Oxamyl, Carbaryl, Carbofuran, Carbosulfan, Methomyl, Methiocarb, Propoxur, Pirimicarb.
- **Pyrethroids:** Allethrins, Deltametrin, Resmethrin, Cypermethrin, Permethrin, Fenvalerate, Pyrethrum.
- **Neonicotinoids:** Acetamiprid, Imidacloprid, Nitenpyram, Thiamethoxam.
- **Organotin compounds**: Triphenyltin acetate, Trivenyltin chloride, Tricyclohexyltin hydroxide, Azocyclotin.
- **Organomercurial compounds:** Ethyl mercuric chloride, Phenyl mercuric bromide.

Pesticide Distribution in Environment

Some of the pesticides listed above are also persistent organic pollutants (POPs) and are discussed further below. Some pesticides are also associated with heavy metal contamination of soils. The recent report by the Intergovernmental Technical Panel on Soils (ITPS) on the impact of plant protection products on soil functions and ecosystem services highlighted the severe impact of copper-based fungicides on earthworms and microbial biomass. These fungicides are widely used in organic viticulture to control vine fungal diseases (FAO and ITPS, 2017). Pesticide persistence, behaviour and mobility are also extremely varied as are the mechanisms involed in their degradation and retention in soils. For example, sorption–desorption, volatilization, chemical and biological degradation, uptake by plants and leaching.

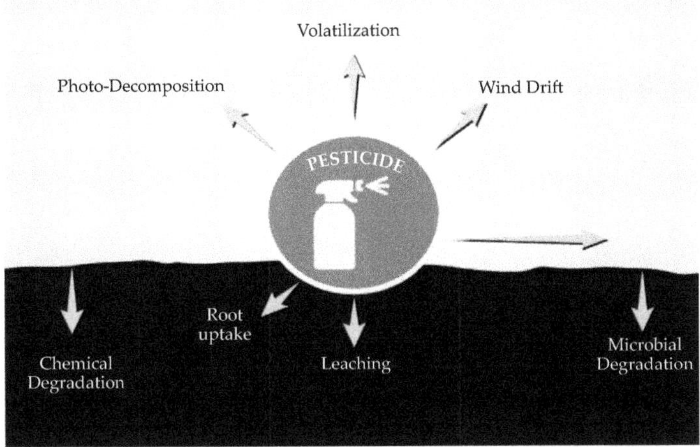

Figure 16.2. Behavior of pesticides in the environment

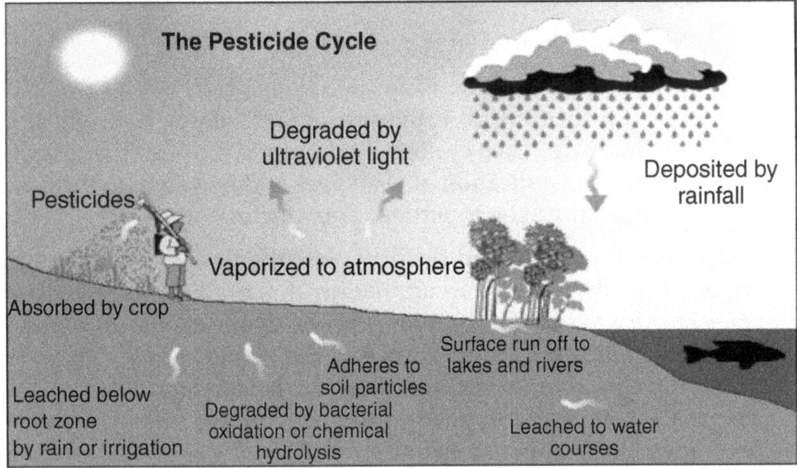

Figure 16.3. Distribution of pesticides in environment

Effects of Chemical Fertilizers and Pesticides

1. **Loss of biodiversity:** The total biodiversity of an area is decreased when pesticides are used, either they are target organisms or non-target organisms. The use of fertilizers lead to the accumulation of chemicals in the soil which leads to the alteration of the pH and ultimately soil organisms either die or they are forced to migrate.

2. **Effect on pollination:** Pesticide contamination poses significant risks to the environment and non-target organisms ranging from beneficial soil microorganisms, to insects, plants, fish, and birds. Honeybees and butterflies play a major role in pollination are very sensitive to pesticides. Due to reduced population of these organisms, pollination is also decreased. This leads to reduced seed and fruit production, which in long run effects the ecology of an area and also cause economic losses.

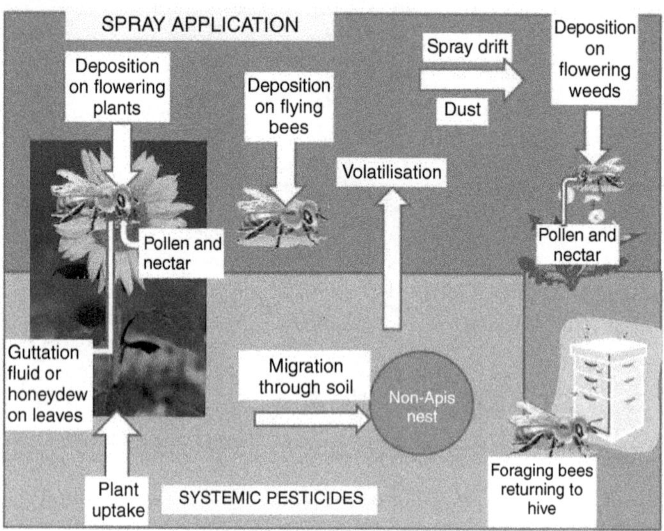

Figure 16.4. Effect of pesticides on honey bees

3. **Effect on nutrient cycling:** With the wide spread use of pesticides, the useful organisms present in the soil are also destroyed, which play an important role in nutrient cycling. For example, nitrogen fixing bacteria are killed due to the application of pesticides and they help in the conversion of atmospheric nitrogen to nitrates and nitrites.

4. **Effect on soil erosion and fertility:** Soil, the basic need of farming may happen to pollute by the accumulation of various heavy metals, through emissions by industries, mining process, disposal of high metal wastes, gasoline, application of fertilizers, sewage sludge, pesticides, wastewater irrigation, coal combustion residues, etc. Historically, a large amount of chemicals is annually applied at the agricultural soils as fertilizers and pesticides. Such applications may result in the increase level of heavy metals, particularly Cd, Pb, and As in the soil. Usage of pesticides, insecticides and other various chemicals in agriculture is very easy, quick and inexpensive solution for controlling weeds and insect pests. However, use of chemicals comes with a significant cost. They have contaminated almost every part of our environment and their residues are found in soil, water, land and air.

5. **Effect on water quality:** Excessive use of fertilizers may pollute the underground water with nitrate and it is so much hazardous to humans or livestock. Nitrate concentrated water can immobilize some of hemoglobin in blood. The most commonly encountered organochlorine pesticides in surface water were dieldrin, heptachlor epoxide, isomers of hexachlorocyclohexane and DDT.

6. **Effect on birds:** Birds are exposed to pesticides through grain feeding, plant feeding and insect eating. Mainly eagles, hawks and owls are affected. Once, they enter the body of birds they effect reproduction and nervous system of the birds. For example, the use of DDT in the fields lead to the no egg shell formation in birds.

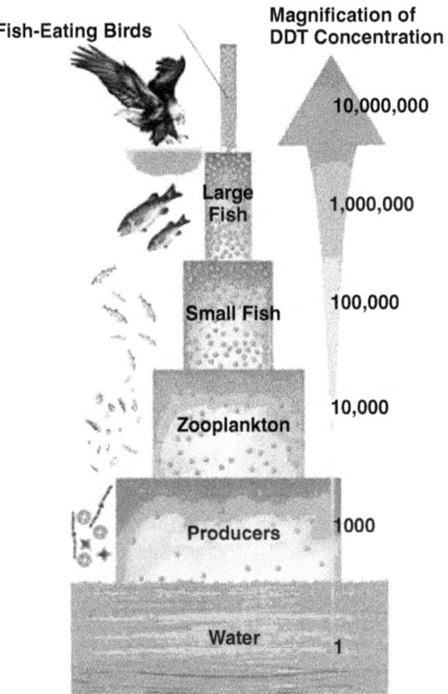

Figure 16.5. Process of Biological Magnification

7. **Disturbance in prey-predator relationship:** Natural prey and predator relationships are severely effect due to the use of pesticides. If prey is killed, then there is a shortage of food for the predator and predator is killed, then the prey population is increased uncontrollably.

8. **Effects on human health:** It has been observed that agrochemicals are causing serious hazards and certain pesticides may affect the human endocrine and immune systems and may promote the development of cancer. Pesticides exposure is through inhalation, contaminated food and dermal exposure which can adversely affect their eyes, skin and the respiratory system.

Moreover, as pesticides are applied over the vegetable which directly enters human or livestock bodies. Organophosphate pesticides have increased in application, because they are both less persistent and harmful for environment than organochlorin pesticides. But, they are associated with acute health problems, such as abdominal pain, dizziness, headaches, nausea, vomiting, as well as skin and eye problems.

There have been many studies intending to establish cancer – pesticides association. These pollutants in water generally are in small quantities, and thus, cannot be seen or tasted. Therefore, their harmful effects do not manifest in humans for several years but led to the escalation of deadly disease like chronic kidney disease. In term of human health, DDT is the cause of many kinds of cancer, acute and persistent injury to the nervous system, lung damage, injury to the reproductive organs, dysfunction of the immune and endocrine systems, birth defects.

Pest Management: A Solution

Large amounts of pesticides are inappropriately used by these farmers, leading to several human health disease, polluting our air, land and water. Since about major proportion of the population relies on agriculture for subsistence, the pesticides are used very widely in agricultural field to increase the production by protecting the yields from potential threat. To safeguard human life and environment from the toxic effects of pesticides, adequate steps need to be taken. Now it is a well established fact that there is the foremost need to step forward towards our mother earth by nurturing it by going for the organic farming system. An answer to this havoc is the organic farming, an environmentally friendly agricultural approach which ultimately leads to proper human health. Moving back to our ancestor's course by performing organic agriculture is a step towards sustainability. Organic agriculture is a holistic production and management system which is supportive of the environment, health and sustainability.

For the control of pest, integrated approach to pest management can be used:

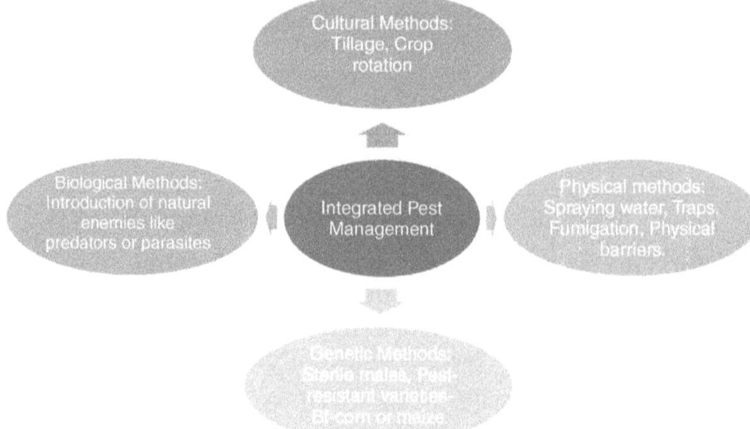

Figure 16.6. Strategies for Integrated Pest Management

Questions

Short Answer Type Questions

1. Define
   - (a) Insecticides
   - (b) Fertilizers
   - (c) Pesticides
   - (d) Biomagnification
2. Name one organochlorine pesticide.

Essay Type Questions

1. Explain the types of pesticides.
2. Describe the fate of pesticides in environment.
3. What are the effects of chemical pesticides on environment?
4. Explain integrated pest management.

Water Pollution

Introduction

Water is one of the renewable resources essential for sustaining all forms of life, food production, economic development, and for general well-being. It is impossible to substitute for most of its uses, difficult to depollute, expensive to transport, and it is truly a unique gift to mankind from nature. Water is also one of the most manageable natural resources as it is capable of diversion, transport, storage, and recycling. All these properties impart to water its great utility for human beings. The surface water and groundwater resources of the country play a major role in agriculture, hydropower generation, livestock production, industrial activities, forestry, fisheries, navigation, recreational activities, etc.

In the last few decades, there has been a tremendous increase in the demand for freshwater due to rapid growth of population and the accelerated pace of industrialization. Human health is threatened by most of the agricultural development activities particularly in relation to excessive application of fertilizers and unsanitary conditions. Anthropogenic activities related to extensive urbanization, agricultural practices, industrialization, and population expansion have led to water quality deterioration in many parts of the world. In addition, deficient water resources have increasingly restrained water pollution control and water quality improvement. Water pollution has been a research focus for government and scientists. Therefore, protecting river water quality is extremely urgent because of serious water pollution and global scarcity of water resources.

Sources of Water Pollution

Water pollution can occur from two sources- 1. Point source, and 2. Non-point source (Table 17.1).

Point sources of pollution are those which have direct identifiable source. Example includes pipe attached to a factory, oil spill from a tanker, effluents coming out from industries. Point sources of pollution include wastewater effluent (both municipal and industrial) and storm sewer discharge and affect mostly the area near it.

Whereas **non-point sources** of pollution are those which arrive from different sources of origin and number of ways by which contaminants enter groundwater or surface water and arrive in the environment from different non-identifiable sources. Examples are runoff from agricultural fields, urban waste, etc. Sometimes pollution that enters the environment in one place has an effect hundreds or even thousands of miles away. This is known as transboundary pollution. One example is the radioactive waste that travels through the oceans from nuclear reprocessing plants to nearby countries. Water pollutants may be i) Organic, and ii) Inorganic water pollutant.

1. **Organic water pollutants:** They comprise of insecticides and herbicides, organohalides and other forms of chemicals; bacteria from sewage and livestock farming; food processing wastes; pathogens; volatile organic compounds, etc.
2. **Inorganic water pollutants:** They may arise from heavy metals from acid mine drainage; silt from surface run-off, logging, slash and burning practices and land filling; fertilizers from agricultural run-off which include nitrates and phosphates, etc. and chemical waste from industrial effluents.

Table 17.1. Characteristics of Point and Non-point Sources of Chemical Inputs to Receiving Waters

Point Sources	Non-point Sources
• Wastewater effluent (municipal and industrial) • Runoff and leachate from waste disposal sites • Runoff and infiltration from animal feedlots • Runoff from mines, oil fields, unsewered industrial sites • Storm sewer outfalls from cities with a population >100,000 • Overflows of combined storm and sanitary sewers • Runoff from construction sites >2 ha	• Runoff from agriculture (including return flow from irrigated agriculture) • Runoff from pasture and range • Urban runoff unsewered and sewered areas with a population <100,000 • Septic tank leachate and runoff from failed septic systems • Runoff from construction sites • Runoff from abandoned mines • Atmospheric deposition over a water surface • Activities on land that generate contaminants, such as logging, wetland conversion, construction, and development of land or waterways

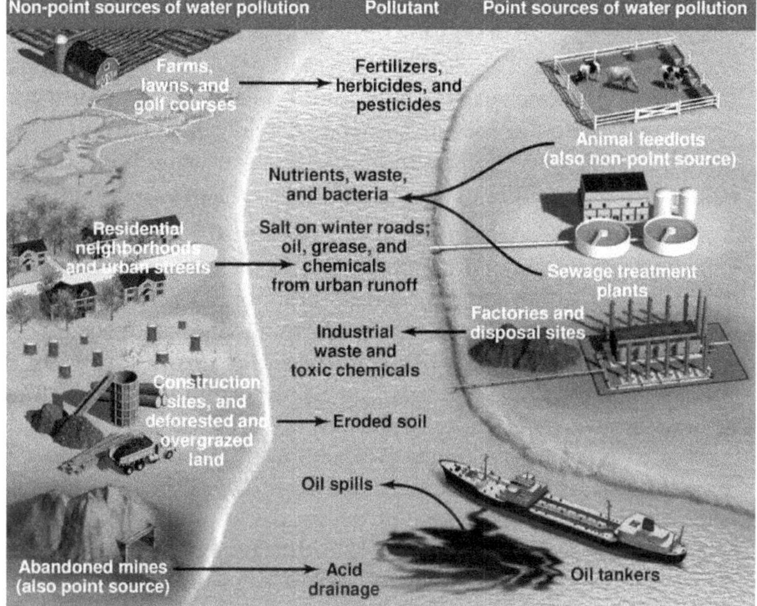

Figure 17.1. Sources of water pollution

Some of the important sources of water pollution are discussed below:

1. **Urbanization:** Urbanization generally leads to higher phosphorus concentrations in urban catchments. Increasing imperviousness, increased runoff from urbanized surfaces, and increased municipal and industrial discharges all result in increased loadings of nutrients to urban streams. This makes urbanization second only to agriculture as the major cause of stream impairment.

2. **Sewage and other oxygen demanding wastes:** Management of solid waste is not successful due to huge volumes of organic and non-biodegradable wastes generated daily. As a consequence, garbage in most parts of India is unscientifically disposed and ultimately leads to increase in the pollutant load of surface and groundwater courses. Sewage can be a fertilizer as it releases important nutrients to the environment such as nitrogen and phosphorus which plants and animals need for growth. Chemical fertilizers used by farmers also add nutrients to the soil, which drain into rivers and seas and add to the fertilizing effect of the sewage. Together, sewage and fertilizers can cause a massive increase in the growth of algae or plankton that facilitate huge areas of oceans, lakes, or rivers creating a condition known as algal bloom thereby reducing the dissolved oxygen content of water and killing other forms of life like fish.

3. **Industrial wastes**: Many of the industries are situated along the banks of river such as steel and paper industries for their requirement of huge amounts of water in manufacturing processes and finally their wastes containing acids, alkalies, dyes and other chemicals are dumped and poured down into rivers

as effluents. Chemical industries concerning with manufacture of Aluminium release large amount of fluoride through their emissions to air and effluents to water bodies. Fertilizer industries generate huge amount of ammonia whereas steel plants generate cyanide. Chromium salts are used in industrial process for the production of sodium dichromate and other compounds containing chromium. All such discharges finally arrive at water bodies in the form of effluents affecting human health and the organism living there.

4. **Agro-chemical wastes:** In the agricultural sector, water and electricity for irrigation are subsidized for political reasons. This leads to wasteful flood irrigation rather than adoption of more optimal practices such as sprinkler and drip irrigation. Cropping patterns and farming practices also do not necessarily encourage the judicious use of water. There are losses of water due to breaches and seepage resulting in waterlogging and salinity. Agro-chemical wastes include fertilizers, pesticides which may be herbicides and insecticides widely used in crop fields to enhance productivity. Improper disposal of pesticides from field farms and agricultural activities contributes a lot of pollutants to water bodies and soils. Some of the pesticides are: DDT, Aldrin, Dieldrin, Malathion, Hexachloro Benzene, etc. Anthropogenic sources of contaminants are contributed from agriculture, domestic and industrial wastes. Nutrient concentrations in streams and rivers have been strongly correlated with human land use and disturbance gradients. Both N and P enrichment have links with the agricultural and urban land uses in the watershed.

5. **Thermal pollution:** Changes in water temperature adversely affect water quality and aquatic biota. Majority of the thermal pollution in water is caused due to human activities. Some of the important sources of thermal pollution are nuclear power and electric power plants, petroleum refineries, steel melting factories, coal fire power plant, boiler from industries which release large amount of heat to the water bodies leading to change in the physical, chemical and biological characteristics of the receiving water bodies. High temperature declines the oxygen content of water; disturbs the reproductive cycles, respiratory and digestive rates and other physiological changes causing difficulties for the aquatic life.

6. **Oil spillage**: Oil discharge into the surface of sea by way of accident or leakage from cargo tankers carrying petrol, diesel and their derivatives pollute sea water to a great extent. Exploration of oil from offshore also lead to oil pollution in water. The residual oil spreads over the water surface forming a thin layer of water-in-oil emulsion.

7. **The disruption of sediments:** Construction of dams for hydroelectric power or water reservoirs can reduce the sediment flow affecting adversely the formation of beaches, increases coastal erosion and reduces the flow of nutrients from rivers into seas (potentially reducing coastal fish stocks). Increased sediment flow can also create a problem. During construction work, soil, rock, and other fine powders sometimes enter nearby rivers in large quantities, causing water to become turbid (muddy or silted). The extra sediment can block the gills of fish, causing them suffocation.

8. **Acid rain pollution:** Water pollution that alters a plant's surrounding pH level, such as due to acid rain, can harm or kill the plant. Atmospheric sulfur

dioxide and nitrogen dioxide emitted from natural and human-made sources like volcanic activity and burning fossil fuels interact with atmospheric chemicals, including hydrogen and oxygen, to form sulfuric and nitric acids in the air. These acids fall down to earth through precipitation in the form of rain or snow. Once acid rain reaches the ground, it flows into waterways that carry its acidic compounds into water bodies. Acid rain that collects in aquatic environments lowers water pH levels and affects the aquatic biota.

9. **Radioactive waste:** Radioactive pollution is caused by the presence of radioactive materials in water. They are classified as small doses which temporarily stimulate the metabolism and large doses which gradually damage the organism causing genetic mutation. Source may be from radioactive sediment, waters used in nuclear atomic plants, radioactive minerals exploitation, nuclear power plants and use of radioisotopes in medical and research purposes.
10. **Introduction of alien species:** In some parts of the world, alien species also known as invasive species are a major problem of water pollution. Outside their normal environment, they have no natural predators, so they rapidly spread and dominate the animals or plants that thrive there. The Mediterranean Sea has been invaded by a kind of alien algae called *Caulerpa taxifolia*. In the Black Sea, an alien jellyfish called *Mnemiopsis leidyi* reduced fish stocks by 90 per cent after arriving in ballast water.
11. **Climate change**: Global warming has also an impact on water resources through enhanced evaporation, geographical changes in precipitation intensity, duration and frequency (together affecting the average runoff), soil moisture, and the frequency and severity of droughts and floods. Future projections using climate models pointed out that there will be an increase in the monsoon rainfall in most parts of India, with increasing greenhouse gases and sulphate aerosols. Relatively small climatic changes can have huge impact on water resources, particularly in arid and semiarid regions such as North-West India. This will have impacts on agriculture, drinking water, and on generation of hydroelectric power, resulting in limited water supply and land degradation.

Effect of Water Pollution on Human Health

Chemicals in water that affect human health: Some of the chemicals affecting human health are the presence of heavy metals such as Fluoride, Arsenic, Lead, Cadmium, Mercury, petrochemicals, chlorinated solvents, pesticides and nitrates. Fluoride in water is essential for protection against dental carries and weakening of the bones. Concentration below 0.5 mg/l causes dental carries and mottling of teeth but exposure to higher levels above 0.5 mg/l for 5-6 years may lead to adverse effect on human health leading to a condition called fluorosis.

Arsenic is a very toxic chemical that reaches the water naturally or from wastewater of tanneries, ceramic industry, chemical factories and from insecticides such as lead arsenate, effluents from fertilizers factories and from fumes coming out from burning of coal and petroleum. Arsenic is highly dangerous for human health causing respiratory cancer, arsenic skin lesion from contaminated drinking water in some districts of West Bengal. Long exposure leads to bladder and lungs cancer.

Lead is contaminated in the drinking water source from pipes, fitting, solder, household plumbing systems. In the human beings, it affects the blood, central nervous system and the kidneys. Child and pregnant women are mostly prone to lead exposure.

Mercury is used in industries such as smelters, manufactures of batteries, thermometers, pesticides, fungicides, etc. The best known example of Mercury pollution in the oceans took place in 1938 when a Japanese factory discharged a significant amount of mercury into Minamata Bay, by contaminating the fish stocks there. It took several years to show its effects. By that time, many local people had eaten the fish and around 2000 were poisoned, hundreds of people were left dead and disabled and the cause for death was named as "Minamata disease" due to consumption of fish containing methyl mercury. It causes chromosomal aberrations and neurological damages to human. Mercury shows biological magnification in aquatic ecosystems.

Cadmium reaches human body through food crop from soil irrigated by affected effluents. Long-term consumption of rice from affected fields by the people living in areas contaminated by cadmium in regions of Japan, resulted into many renal vdiseases like "itai-itai disease", nephritis and nephrosis.

Waterborne disease: Microorganisms play a major role in water quality and the microorganisms that are concerned with water borne diseases are *Salmonella* sp., *Shigella* sp., *Escherichia coli* and *Vibrio cholera*. All these cause typhoid fever, diarrhoea, dysentery, gastroenteritis and cholera. The most dangerous form of water pollution occurs when faeces enter the water supply. Many diseases are perpetuated by the faecal-oral route of transmission in which the pathogens are shed only in human faeces. Presence of faecal coliforms of *E. coli* is used as an indicator for the presence of any of these water borne pathogens. In recent years, the widespread reports of pollutants in groundwater have increased public concern about the quality of groundwater. Children are generally more vulnerable to intestinal pathogens and it has been reported that about 1.1 million children die every year due to diarrhoeal diseases.

Effect of Water Pollution on Plants

The following are the effects of water pollution on plants:

1. **Effects of acid deposition:** Many of the gases from acid, aerosols and other acidic substances released into the atmosphere from industrial or domestic sources of combustion from fossil fuels finally fall down to ground and reach the water bodies along with run-off rainwater from polluted soil surfaces thereby causing acidification of water bodies by lowering its pH. In many countries chemical substances like sulphates, nitrates and chloride have been reported to make water bodies such as lakes, river and ponds acidic.

2. **Nutrient deficiency in aquatic ecosystem:** Population of decomposing microorganisms like bacteria and fungi decline in acidified water which in turn reduces the rate of decomposition of organic matter affecting the nutrient cycling. The critical pH for most of the aquatic species is 6.0. The diversity of species decline below this pH whereas the number and abundance of acid tolerant species increases. Proliferation of filamentous algae rapidly forms a thick mat at the initial phase of the acidification of water. Diatoms and green

algae disappear below pH 5.8. *Cladophora* is highly acid tolerant species and is abundant in acidic freshwater bodies. Macrophytes are generally absent in acidic water as their roots are generally affected in such water resulting in poor plant growth. *Potamogeton pectinalis* is found in acidified water. It is observed that plants with deep roots and rhizomes are less affected while plants with short root systems are severely affected in acidic water.

3. **Effects of organic matter deposition:** Organic matter from dead and decaying materials of plants and animals is deposited directly from sewage discharges and washed along with rainwater into water bodies causing increase in decomposers / microbes such as aerobic and anaerobic bacteria. Rapid decomposition of organic matter increase nutrient availability in water favouring the luxuriant growth of planktonic green and blue-green algal bloom. In addition many of the macrophytes like *Salvinia, Azolla, Eicchhornia,* etc. grow rapidly causing reduced penetration of light into deeper layer of water body with gradual decline of the submerged flora. This condition results in reducing the dissolved oxygen and increase in the biological oxygen demand (B.O.D). The B.O.D of unpolluted fresh water is usually below 1mg/l while that of organic matter polluted water is more than 400 mg/l.

4. **Effects of detergent deposition:** Detergents from domestic and industrial uses wash down into water bodies causing serious effects on plants. Detergents contain high phosphates which results in phosphate-enrichment of water. Phosphates enter the plants through roots or surface absorption causing retarded growth of plants, elongation of roots, carbon dioxide fixation, photosynthesis, cation uptake, pollen germination and growth of pollen tubes, destruction of chlorophylls and cell membranes and denaturation of proteins causing enzyme inhibition in various metabolic processes.

5. **Effects of agricultural chemicals:** Chemicals from fertilizers, pesticides, insecticides, herbicides, etc. applied to crops in excess are washed away with rainwater as runoff, then enter soil and finally arrive at the water bodies. Chemicals from fertilizers result in eutrophication by enrichments of nutrients. Ammonium from fertilizers is acidic in nature causing acidification of water. Similarly pesticides, herbicides and insecticides also cause change in pH of the water bodies. Most common effect of these substances is the reduction in photosynthetic rate. Some may uncouple oxidative phosphorylation or inhibit nitrate reductase enzyme. The uptake and bioaccumulation capacities of these substances are great in macrophytic plants due to their low solubility in water.

6. **Effects of industrial wastes:** Effluents from industries contain various organic and inorganic waste products. Fly ash form thick floating cover over the water thereby reducing the penetration of light into deeper layers of water bodies. Fly ash increases the alkalinity of water and cause reduced uptake of essential bases leading to death of aquatic plants. Liquid organic effluents change the pH of water and the specific toxicity effects on the aquatic plants vary depending on their chemical composition. There may be synergistic, additive or antagonistic interactions between metals with

respect to their effects on plants however these effects are reduced in hard and buffered freshwater bodies.

7. **Effects of silt deposition:** Deposition of silt in water bodies occurs as a result of erosion carrying silt laden water and due to flood. It increases the turbidity of water and reduces light penetration in deep water causing decline in abundance of submerged plants. Siltation inhibits the growth of aquatic plants. Abundance of phytoplankton is affected due to reduction in surface exchange of gases and nutrients. Plants that are tolerant to turbidity are abundant followed by those which are intermediate and the least tolerant species. Plants such as *Polygonum, Sagittaria,* etc. are found to grow in dominance.

8. **Effects of oil spillage:** Oil pollution due to spillage of oil tankers and storage containers prevents oxygenation of water and depletes the oxygen content of the water body by reducing light transmission inhibiting the growth of planktons and photosynthesis in macrophytes.

9. **Effects of thermal pollution:** The release of heated water into water bodies from the thermal power plants has an adverse effect on the aquatic life. It reduces the activity of aerobic decomposers due to oxygen depletion because of high temperature. With decreased organic matter decomposition, the availability of nutrients in the water bodies is jeopardised. Aquatic plants show reduced photosynthesis rate due to inhibition of enzyme activity with increased temperature. Primary productivity and diversity of aquatic plant species decline because of increased temperature of water bodies as a result of thermal pollution.

10. **Effect of nutrient enrinchement**: Nutrient enrichment in aquatic water bodies leads to eutrophication which is a process whereby water bodies receive excess inorganic nutrients, especially N and P, stimulating excessive growth of plants and algae. Eutrophication can happen naturally in the course of normal succession of some freshwater ecosystems. However, when the nutrient enrichment is due to the activities of humans, it is referred to as "cultural eutrophication", where the rate of nutrient enrichment is greatly intensified. Nitrogen and phosphorus, in particular, encourage growth because they stimulate photosynthesis. This is why they are common ingredients in plant fertilizers. When agricultural runoff pollutes waterways with nitrogen and phosphorus rich fertilizers, the nutrient-enriched waters often paves way to algal bloom leading to eutrophication. The result in oxygen depletion and dying of fishes due to suffocation.

11. **Phytotoxicity effects on plants:** When chemical pollutants build up in aquatic or terrestrial environments, plants can absorb these chemicals through their roots. *Phytotoxicity* occurs when toxic chemicals poison plants. The symptoms of phytotoxicity on plants include poor growth, dying seedlings and dead spots on leaves. For example, mercury poisoning which many people associate with fish can also affect aquatic plants, as mercury compounds build up in plant roots and bodies result in bioaccumulation. As animals feed on polluted food the increasing levels of mercury is built up through food chain.

Control of Water Pollution

The key challenges to better management of the water quality in India comprise of temporal and spatial variation of rainfall, uneven geographic distribution of surface water resources, persistent droughts, overuse of groundwater and contamination, drainage and salinisation and water quality problems due to treated, partially treated and untreated wastewater from urban settlements, industrial establishments and runoff from irrigation sector besides poor management of municipal solid waste and animal dung in rural areas (CPCB Report, 2013). Some of the control measures are given below:

- The Ganga Action Plan and the National River Action Plan are being implemented for addressing the task of trapping, diversion and treatment of municipal wastewater.
- In most parts of the country, waste water from domestic sources is hardly treated, due to inadequate sanitation facilities. This waste water, containing highly organic pollutant load, finds its way into surface and groundwater courses near the vicinity of human habitation from where further water is drawn for use. Considerable investments should be done to install the treatment systems.
- With rapid industrialization and urbanization, the water requirement for energy and industrial use is estimated to rise to about 18 per cent (191 bcm) of the total requirements in 2025 (CPCB Report, 2013). Poor environmental management systems, especially in industries such as thermal power stations, chemicals, metals and minerals, leather processing and sugar mills, have led to discharge of highly toxic and organic wastewater. This has resulted in pollution of the surface and groundwater sources from which water is also drawn for irrigation and domestic purpose. The enforcement of regulations regarding discharge of industrial wastewater and limits to extraction of groundwater needs to be considerably strengthened, while more incentives are required for promoting wastewater reuse and recycling.
- For the agricultural sector, water and electricity for irrigation are subsidized for political reasons. This leads to wasteful flood irrigation rather than adoption of more optimal practices such as sprinkler and drip irrigation. Optimized irrigation, cropping patterns and farming practices should be encouraged for judicious use of water.
- The water quality management in India is accomplished under the provision of Water (Prevention and Control of Pollution) Act, 1974 that was amended in 1988. The basic objective of this Act is to maintain and restore the wholesomeness of national aquatic resources by prevention and control of pollution. The Water (Prevention and Control of Pollution) Cess Act was enacted in 1977, to provide for the levy and collection of a cess on water consumed by persons operating and carrying on certain types of industrial activities.
- The Central Pollution Control Board (CPCB) has established a network of monitoring stations on aquatic resources across the country. The water quality monitoring and its management are governed at state/union territory level in India. The network covers 28 states and 6 Union Territories (CPCB Report, 2013). Water quality monitoring is therefore an imperative

prerequisite in order to assess the extent of maintenance and restoration of water bodies.

- ☆ There should be ban on washing of clothes and laundry alongside the river bank.
- ☆ Industries should install Effluent Treatment Plant (ETP) to control the pollution at source.
- ☆ All towns and cities must have Sewage Treatment Plants (STPs) that clean up the sewage effluents.
- ☆ Improper use of fertilizers, herbicides and pesticides in farming should be stopped and organic methods of farming should be adoped. Cropping practices in riparian zone should be banned to protect the riparian vegetation growing there.
- ☆ Religious practices that pollute river water by dumping colourful paints of idols containing harmful synthetic chemicals should be stopped.
- ☆ Rainwater harvesting should be practiced to prevent the depletion of water table.
- ☆ Making people aware of the problem is the first step to prevent water pollution. Hence, importance of water and pollution prevention measures should be a part of awareness and education programme.
- ☆ Polluter pays principle should be adopted so that the polluters will be the first people to suffer by way of paying the cost for the pollution. Ultimately, the polluter pays principle should be designed to prevent people from polluting and making them behave in an environmentally responsible manner.
- ☆ As riparian vegetation helps in making the river water clean because of the multiple functions, to prevent people from felling and clearing down of riparian forest zones for road construction, agricultural practices, recreational and tourism, sand mining, quarrying and clay mining, etc. community should play a regulatory role.

Effluent Treatment

Ideally no effluent should be released into the environment before it has received adequate and satisfactory treatment. Treatment and disposal of Sewage is important at it can harm the aquatic environment adversely. Sewage is a turbid liquid, consisting of 99.9% water containing a complex of organic and inorganic matter, in the form of suspended solids, colloidal particles, dissolved compounds and microorganisms such as Protozoa, bacteria and viruses. The objectionable odour and colour is largely caused by the 66% organic matter present, and the anaerobic bacteria action that takes place within it. The organic matter is present as; paper, faeces, urine, soap, detergents, fats, oils, greases, and food materials. The inorganic substances present include; sand, clay, ammonia and ammonium salts derived from the decomposition of urine, metallic salts, nitrates, phosphates, etc. The precise composition of sewage varies depending on its origin and its industrial effluent input, which may be upto 50% by volume. Modern sewage treatments basically aim to remove the floating and suspended solids, and provide biological treatment of the organic matter present.

Sewage Treatment Can be Divided into Three Stages

1. Preliminary stage 2. Primary stage 3. Secondary stage.

Some treatment plants provide a final tertiary stage to produce a higher quality effluent.

1. **Preliminary stage:** The preliminary treatment removes large suspended debri and solid particles by screening.
2. **Primary Stage:** In this stage sedimentation takes place which results in the removal of some 55% of settleable solids as sludge, and 35% reduction in BOD value. The sewage at this point contains suspended colloids, finely divided solids and dissolved solutes.
3. **Secondary Stage**: The activated sludge method is applied in the modern sewage treatment plants. The primary effluent is agitated and aerated in large tanks, in the presence of flocculent suspension of activated sludge containing microorganisms. In this biological treatment, microorganisms utilize the organic matter and break it down into non-pollutive inorganic compounds. The treatment has been called self-purification, and it takes place naturally in streams and rivers. However, in sewage works, optimum conditions are provided to enable the biodegradation of organic matter to take place more rapidly than in natural water.

Following the secondary biological treatment, there is a further sedimentation phase to allow more settlement of suspended solids. The effluent emerging from the secondary stage can be discharged into water courses. However, it is further recommended that a tertiary stage be incorporated where effluent from secondary stage is discharged into a wet-land for further purification and passed into a fish pond, before release into water courses.

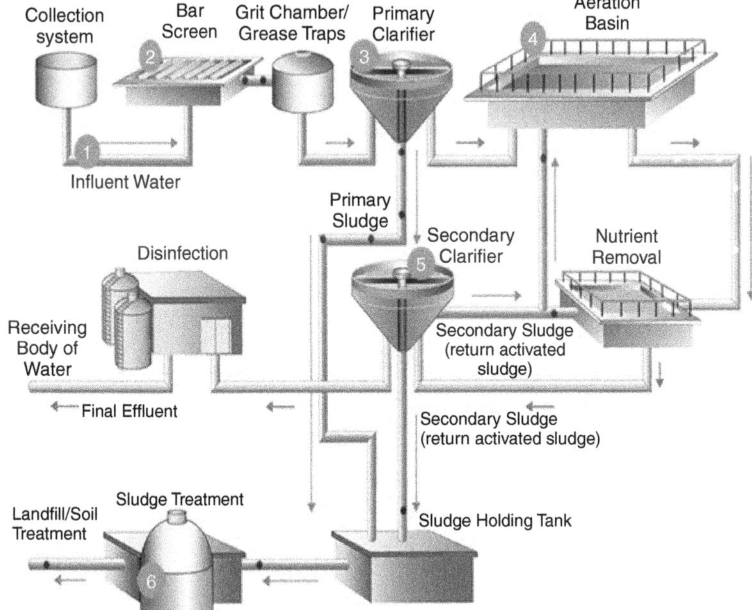

Figure 17.2. Flow diagram of a Sewage Treatment Plant

Tertiary treatment has its limitations, for example it cannot completely compensate for poor secondary treatment and it does not remove all suspended solids, organic matter and dissolved solutes. Consequently the effluent cannot be used as drinking water without the normal chemical treatment given to natural waters. The primary and secondary treatments do not completely remove all potentially polluting materials, especially metals, nitrites, phosphates, ammonium compounds and bacteria.

Questions

Short Answer Type Questions

1. Define water pollution.
2. What is thermal pollution?
3. Define eutrophication.
4. What is domestic sewage?
5. What was the cause of Minamata diseases?

Essay Type Questions

1. Explain the various sources of water pollution.
2. Describe biological magnification and its process.
3. Write down the various effects of water pollution on aquatic ecosystem.
4. What are the different methods for the control of water pollution?
5. What is wastewater treatment? Explain the process of waste water treatment.
6. What are the effects of thermal pollution on aquatic ecosystems?

18

Biodiversity Conservation

A wide variety of living organisms including plants, animals and micro-organisms with which we share with this planet earth, makes the world a beautiful place to live in. Living organisms exist almost everywhere from mountain peaks to the ocean depths; from deserts to the rainforests. They vary in their habit and behavior, shapes, sizes and color. The remarkable diversity of living organisms form an inseparable and significant parts of our planet however, the ever increasing human population is posing serious threats to biodiversity.

Sum total of all the variety of living organisms on earth constitute biodiversity. Biological diversity is usually considered at three different levels – a) genetic diversity, i.e., at genetic level, b) species diversity, i.e., at the level of species, and c) ecosystem diversity, i.e., at the level of ecosystem.

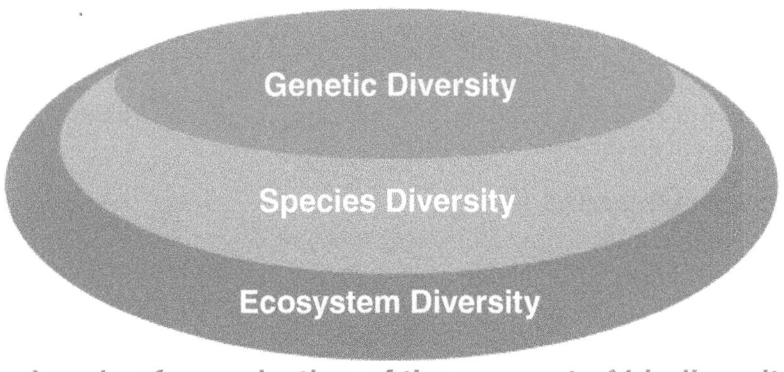

Levels of organization of the concept of biodiversity

Figure 18.1. Levels of Biodiversity

Genetic Diversity

Each species, varying from bacteria to higher plants and animals, stores an immense amount of genetic information. This variation of genes, not only of numbers but of structure also, is of great value as it enables a population to adapt to its environment and to respond to the process of natural selection. If a species has more genetic variation, it can adapt better to the changed environmental conditions. Lower diversity in a species leads to genetic uniformity of genetically similar crop plants. This homogeneity is desirable in producing uniform quality of grain. But genetic uniformity restricts adaptability of a species to environmental stress as all the plants have same level of resistance.

With the above background, genetic diversity refers to the variety of genes contained within species of plants, animals and micro-organisms. New genetic variation in individuals occurs by gene and chromosomal mutation, and in organisms with sexual reproduction may be spread across the population by recombination. The amount of genetic variation (gene pool) present in an inter-breeding population is shaped or decided by the process of natural selection. Selection leads to certain genetic attributes being preferred and results in changes in the frequency of genes within this pool. This forms the basis of adaptation among the living organisms.

Species Diversity

Species diversity refers to the variety of species within a geographical area. Species diversity can be measured in terms of:

(a) **Species richness** – refers to the number of various species in a defined area.

(b) **Species abundance** – refers to the relative numbers among species. For example, the number of species of plants, animals and micro-organisms may be more in an area than that recorded in another area.

(c) **Taxonomic or phylogenetic diversity** – refers to the genetic relationships between different groups of species.

Kinds of species that are present in an area are also important. When taxonomically unrelated species are present in an area, the area represents greater species diversity as compared to an area represented by taxonomically related species.

In addition, biodiversity in land ecosystems generally decreases with increasing altitude. The other factors that influence biodiversity are amount of rainfall and nutrient level in soil. In marine ecosystems, species richness tends to be much higher in continental shelves.

Ecosystem Diversity

It refers to the presence of different types of ecosystems. For instance, the tropical south India with rich species diversity will have altogether different structure compared to the desert ecosystem which has far less number of plant and animal species. Likewise, the marine ecosystem although has many types of fishes, yet it differs from the freshwater ecosystem of rivers and lakes in terms of its characteristics. So such variations at ecosystem level are termed as ecosystem diversity.

As stated above, ecosystem diversity encompasses the broad differences between ecosystem, and the diversity of the habitats and ecological processes occurring within each ecosystem type. India has very diverse terrestrial and aquatic ecosystems ranging from ice-capped Himalayas to deserts, from arid scrub to grassland to wetlands and tropical rainforests, from coral reefs to the deep sea. Each of these comprises a great variety of habitats and interactions between and within biotic and abiotic components. The most diversity-rich are western-ghats and the north-eastern region. A very large number of species found in these ecosystems are endemic or found in these areas only in India, i.e., they are found nowhere else except in India.

Hotspots of Biodiversity

Biodiversity is not uniformly distributed across the geographical regions of the earth. Certain regions of the world are very rich in biodiversity. We call such areas as "mega diversity zones". We also refer to them as "hotspots". For example, India accounts for only 2.4 % of the land area of the world; but it contributes approximately 8% species to the global diversity due to existence of such pockets.

Norman Myers, a British Ecologist, developed the concept of hotspots in 1988 to designate priority areas for *in situ* conservation. According to him, the hotspots are the richest and the most threatened reservoirs of biodiversity on the earth. The criteria for determining a hotspot are:

I) The area should support >1500 endemic species.

Ii) It must have lost over 70 % of the original habitat.

This idea of identifying hotspots was put forth by Norman Myers in 1988. By now, a total of 35 biodiversity hotspots have been identified out of which most of them lie in tropical forests. Almost 2.3% of the land surface of Earth is represented by these hotspots. These also comprise of around 50% of the world's most common plant species and 42% of terrestrial vertebrates prevalent. Sadly, these biodiversity hotspots have been losing 86% of their habitats some of which are still on the verge of extinction due to serious threats posed by climate change and human intervention.

Out of 35 hotspots, 4 hotspots of biodiversity are found in India, namely Himalaya: Includes the entire Indian Himalayan region (and that falling in Pakistan, Tibet, Nepal, Bhutan, China and Myanmar) Indo-Burma: Includes entire North-eastern India, except Assam and Andaman group of Islands (and Myanmar, Thailand, Vietnam, Laos, Cambodia and southern China) zundalands: Includes Nicobar group of Islands (and Indonesia, Malaysia, Singapore, Brunei, Philippines) Western Ghats and Sri Lanka: Includes entire Western Ghats (and Sri Lanka)

Why is Biological Diversity Important?

Humans depend for their sustenance, health, well-being and cultural growth on nature. Biotic resources provide food, fruit, seed, fodder, medicines and a host of other goods and services. The enormous diversity of life is of immense value, imparting resilience to ecosystems and natural processes. Biodiversity also has enormous social and cultural importance.

The Value of Biological Diversity

The various benefits of biological diversity can be grouped under three categories: a) ecosystem services, b) biological resources, and c) social benefits.

Ecosystem services

Living organisms provide many ecological services free of cost that are responsible for maintaining ecosystem health. Thus biodiversity is essential for the maintenance and sustainable utilization of goods and services from ecological system as well as from individual species.

(i) **Protection of water resources:** Natural vegetation cover helps in maintaining hydrological cycles, regulating and stabilizing water run-off and acting as a buffer against extreme events such as floods and droughts. Vegetation removal results in siltation of dams and waterways. Wetlands and forests act as water purifying systems, while mangroves trap silt thereby reducing impacts on marine ecosystems.

(ii) **Soil protection:** Biological diversity helps in the conservation of soil and retention of moisture and nutrients. Clearing large areas of vegetation cover has been often seen to accelerate soil erosion, reduce its productivity and often result in flash floods. Root systems allows penetration of water to the subsoil layer. Root system also brings mineral nutrients to the surface by nutrient uptake.

(iii) **Nutrient storage and cycling:** Ecosystem perform the vital function of recycling nutrients found in the atmosphere as well as in the soil. Plants are able to take up nutrients, and these nutrients then can form the basis of food chains, to be used by a wide range of life forms. Nutrients in the soil, in turn, is replenished by dead or waste matter which is transformed by micro-organisms; this may then feed others such as earthworms which also mix and aerate the soil and make nutrients more readily available.

(iv) **Pollution reduction:** Ecosystems and ecological processes play an important role in maintenance of gaseous composition of the atmosphere, breakdown of wastes and removal of pollutants. Some ecosystems, especially wetlands have the ability to breaking down and absorb pollutants. Natural and artificial wetlands are being used to filter effluents to remove nutrients, heavy metals, suspended solids; reduce the BOD (Biological Oxygen Demand) and destroy harmful micro-organisms. Excessive quantities of pollutants, however, can be detrimental to the integrity of ecosystems and their biota.

(v) **Climate stability:** Vegetation influences climate at macro as well as micro levels. Growing evidence suggests that undisturbed forests help to maintain the rainfall in the vicinity by recycling water vapor at a steady rate back into the atmosphere. Vegetation also exerts moderating influence on micro climate. Cooling effect of vegetation is a common experience which makes living comfortable. Some organisms are dependent on such microclimates for their existence.

(vi) **Maintenance of ecological processes:** Different species of birds and predators help to control insect pests, thus reduce the need and cost of artificial control measures. Birds and nectar-loving insects which roost and breed in natural habitats are important pollinating agents of crop and wild plants. Some habitats protect crucial life stages of wildlife populations such as spawning areas in mangroves and wetlands.

Without ecological services provided by biodiversity it would not be possible to get food, pure air to breathe and would be submerged in the waste produced.

Biological resources of economic importance

(i) **Food, fibre, medicines, fuel wood and ornamental plants:** Five thousand plant species are known to being used as food by humans. Presently about 20 species feed the majority of the world's population and just 3 or 4 only are the major staple crops to majority of population in the world. A large number of plants and animals materials are used for the treatment of various ailments. It is estimated that at least 70% of the country's population rely on herbal medicines and over 7000 species of plants are used for medicinal purposes. Wood is a basic commodity used worldwide for making furniture and for building purposes. Fire wood is the primary source of fuel widely used in third world countries. Wood and bamboo are used for making paper. Plants are the traditional source of fibre such as coir, hemp, flax, cotton and jute.

(ii) **Breeding material for crop improvement:** Wild relatives of cultivated crop plants contain valuable genes that are of immense genetic value in crop improvement programmes. Genetic material or genes of wild crop plants are used to develop new varieties of cultivated crop plants for restructuring of the existing ones for improving yield or resistance of crops plants. For example, rice grown in Asia is protected from four main diseases by genes contributed by a single wild rice variety.

(iii) **Future resources:** There is a clear relationship between the conservation of biological diversity and the discovery of new biological resources. The relatively few developed plant species currently cultivated have had a large amount of research and selective breeding applied to them. Many presently under-utilised food crops have the potential to become important crops in the future. Knowledge of the uses of wild plants by the local people is often a source for ideas on developing new plant products.

Social benefits

(i) **Recreation:** Forests, wildlife, national parks and sanctuaries, garden and aquaria have high entertainment and recreation value. Ecotourism, photography, painting, film-making and literary activities are closely related.

(ii) **Cultural values:** Plants and animals are important part of the cultural life of humans. Human cultures have co-evolved with their environment and biological diversity can be impart a distinct cultural identity to different communities. The natural environment serves the inspirational, aesthetic, spiritual and educational needs of the people, of all cultures.

Research, Education and Monitoring

There is still much to learn on how to get better use from biological resources, how to maintain the genetic base of harvested biological resources, and how to rehabilitate degraded ecosystems. Natural areas provide excellent living laboratories for such studies, for comparison with other areas under systems of use and for valuable research in ecology and evolution.

Uniqueness of Indian Biodiversity

India is uniquely rich in all aspects of biodiversity including ecosystem, species and genetic biodiversity. For any one country in the world, it has perhaps the largest array of environmental situations by virtue of its tropical location, varied physical features and climate types. India has the widest variety of ecosystems. With only 2.4% of the land area, India accounts for 7-8 % of the recorded species of the world. More than 45000 species of plants and 81,000 species of animals are found in India. India is also one of the eight primary centers of origin of cultivated plants and has a rich agricultural biodiversity.

The trans-Himalayan region with its sparse vegetation has the richest wild sheep and goat community in the world. The snow leopard (*Panthera uncia*) and Black-necked Crane (*Grus nigricollis*) are found here. The Great Indian Bustard (*Ardeotis nigriceps*) which is highly endangered bird, is found in (Gujrat) region, rich in extensive grasslands.

North-east India is one of the richest regions of biodiversity in the country. It is especially rich in orchids, bamboos, ferns, citrus, banana, mango and jute.

India is also rich in coral reefs. Major reef formations in Indian seas occur in the Gulf of Mannar, Palk Bay, Gulf of Kutch, the Andaman and Nicobar Islands and the Lakshadweep. The threat to mangroves trees (growing in marshy lands) and coral reefs comes from the biotic pressure such as extraction for market demands, fishing, land-use changes in surrounding areas, and from pollution of water, etc.

Causes of Biodiversity Depletion

Loss of species is a serious cause of concern for human survival. It has been observed that 79 species of mammals, 44 of birds, 15 of reptiles and 3 of amphibians are threatened. Nearly 1500 species of plants are endangered in India. The threat to survival or loss may be caused in the following three ways:

Direct ways : Deforestation, hunting, poaching, commercial exploitation.

Indirect ways : Loss or modification of the natural habitats, introduction of exotic species, pollution, etc.

Natural causes : Climate change.

Among these causes, habitat destruction and overexploitation are the main.

(i) **Habitat** (natural home) destruction may result from clearing and burning forests, draining and filling of wetlands, converting natural areas for agricultural or industrial uses, human settlements, mines, building of roads and other developmental projects. This way the natural habitats of organisms are changed or destroyed. These change either kill or force out many species from the area causing disruption of interactions among the species. Fragmentation of large forest tracts (e.g., the corridores) affects the species occupying the deeper part of the forest and are first to disappear. Apart from the direct loss of species during the development activities, the new environment is unsuitable for the species to survive. Overexploitation reduces the size of the population of a species and may push it towards extinction.

(ii) **Introduction of exotic species:** Seeds catch on people's clothes. Mice, rats and birds hitch-hike on ships. When such species land in new places, they breed extra fast due to absence of any enemy and often wipe out the native species already present there. Exotic species (new species entering geographical region) may wipe out the native ones. A few examples are:

- ***Parthenium hysterophorus*** (Congress grass- a tropical American weed) has invaded many of the vacant areas in cities, towns and villages in India leading to removal of the local plants and the dependent animals.
- **Nile perch**, an exotic predatory fish introduced into Lake Victoria (South Africa) threatened the entire ecosystem of the lake by eliminating several native species of the small Cichlid fish that were endemic to this freshwater aquatic system.
- **Water hyacinth** clogs lakes and riversides and threatens the survival of many aquatic species. This is common in Indian plains.
- ***Lantana camara*** (an American weed) has invaded many forest lands in various parts of India and wiped out the native grass species.

(iii) **Pollution:** Air pollution, acid rain destroy forests. Water pollution kills fishes and other aquatic plants and animals. Toxic and hazardous substances drained into waterways kill aquatic life. Oil spills kill coastal birds, plants and other marine animals. Plastic trash entangles wildlife. It is easy to see how pollution is a big threat to biodiversity.

(iv) **Population growth and poverty:** Over six billion people live on the earth. Each year, 90 million more people are added. All these people use natural resources for food, water, medicine, clothes, shelter and fuel. Need of the poor and often greed of the rich generate continuous pressure resulting in over-exploitation and loss of biodiversity.

The World Conservation Union (IUCN) (formerly known as International Union for the Conservation of Nature and Natural Resources (IUCN) has recognized eight Red List categories according to the conservation status of species.

(i) **Extinct:** A taxon is extinct when there is no reasonable doubt that the last individual has died.

(ii) **Extinct in the wild:** A taxon is extinct in the wild when exhaustive surveys in known and/or expected habitats have failed to record an individual.

(iii) **Critically endangered:** A taxon is critically endangered when it is facing high risk of extinction in the wild in immediate future.

(iv) **Endangered:** A taxon is endangered when it is not critically endangered but is facing a very high risk of extinction in the wild in near future.

(v) **Vulnerable:** A taxon is vulnerable when it is not critically endangered or endangered but is facing high risk of extinction in the wild in the medium term future.

(vi) **Lower risk:** A taxon is lower risk when it has been evaluated and does not satisfy the criteria for critically endangered, endangered or vulnerable.

(vii) **Data deficient:** A taxon is data deficient when there is inadequate information to make any direct or indirect assessment of its risk of extinction.

(viii) **Not evaluated:** A taxon is not evaluated when it has not yet been assessed against the above criteria.

Conservation of Biodiversity

Conservation is the planned management of natural resources, to retain the balance in nature and retain the diversity. It also includes wise use of natural resources in such a way that the needs of present generation are met and at the same time leaving enough for the future generations. Conservation of biodiversity is important to:

- ☆ Prevent the loss of genetic diversity of a species,
- ☆ Save a species from becoming extinct, and
- ☆ Protect ecosystems damage and degradation.

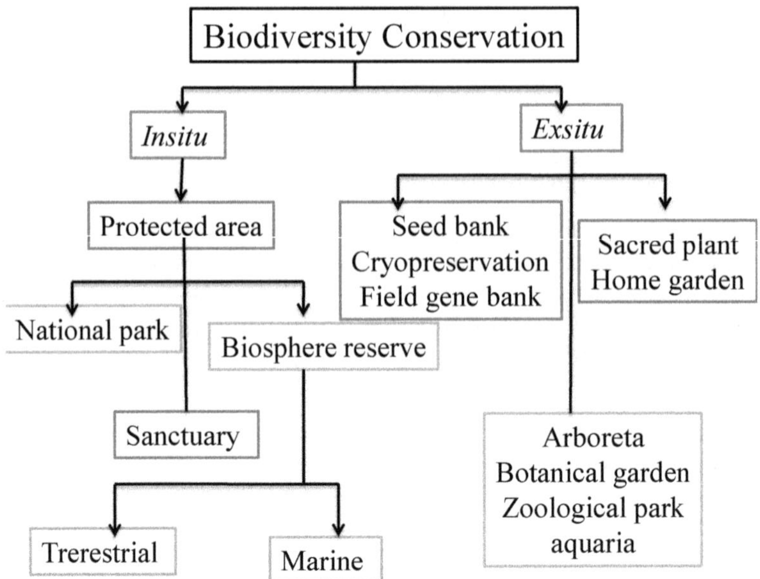

Figure 18.2. Biodiversity Conservation strategies

Conservation Strategies

Conservation efforts can be grouped into the following two categories:

1. ***Insitu*** (on-site) conservation includes the protection of plants and animals within their natural habitats or in protected areas. Protected areas are land or sea dedicated to protect and maintain biodiversity.
2. ***Exsitu*** (off-site) conservation of plants and animals outside their natural habitats. These include botanical gardens, zoo, gene banks, seed bank, tissue culture and cryopreservation.

In Situ Methods

I) Protection of habitat: The main strategy for conservation of species is the protection of habitats in representative ecosystems. Currently, India has ninety six National Parks, five hundred Wildlife Sanctuaries, thirteen Biosphere Reserves, twenty-seven Tiger Reserves and eleven Elephant Reserves covering an area of 15.67 million hectares or 4.7 % of the geographical area of the country. Twenty one wetlands, thirty mangrove areas and four coral reef areas have been identified for intensive conservation and management purposes by the Ministry of Environment and Forests, Govt. of India.

- **National parks and sanctuaries:** India is unique in the richness and diversity of its vegetation and wildlife. India's national parks and wildlife sanctuaries (including bird sanctuaries) are situated Ladakh in Himalayas to Southern tip of Tamil Nadu with its rich biodiversity and heritage. Wildlife sanctuaries in India attract people from all over the world as the rarest of rare species are found here. With 96 national parks and over 500 wildlife sanctuaries, the range and diversity of India's wildlife heritage is unique.
- **Biosphere Reserves:** These are representative parts of natural and cultural landscapes extending over large areas of terrestrial or coastal/marine ecosystems which are internationally recognized within UNESCO's Man and the Biosphere Programme. Thirteen biodiversity- rich representative ecosystems, largely within the forest land (total area – 53,000 sq. km.), have been designated as Biosphere Reserves in India.

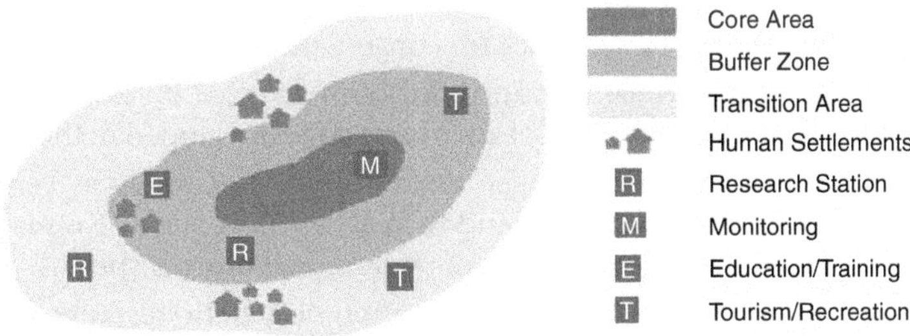

Figure 18.3. Biosphere area zonation

The concept of Biosphere Reserves (BR) was launched in 1975 as a part of UNESCO's Man and Biosphere Programme, dealing with the conservation of ecosystems and the genetic material they contain. A Biosphere Reserve consists of core, buffer and transition zones.

(a) The **core zone** is fully protected and natural area of the Biosphere Reserve least disturbed by human activities. It is legally protected ecosystem in which entry is not allowed except with permission for some special purpose. Destructive sampling for scientific investigations is prohibited.

(b) The **buffer zone** surrounds the core zone and is managed to accommodate a greater variety of resource use strategies, and research and educational activities.

(c) The **transition zone**, the outermost part of the Biosphere Reserve, is an area of active cooperation between the reserve management and the local people, wherein activities like settlements, cropping, forestry , recreation and other economic activities that are in harmony with the conservation goals. Till date, there were 553 biosphere reserves located in 107 countries. The main functions of the biosphere reserves are:

- **Conservation:** Long-term conservation of representatives, landscapes and different types of ecosystems, along with all their species and genetic resources.
- **Development:** Encourages traditional resource use and promote economic development which is culturally, socially and ecologically sustainable.
- **Scientific research, monitoring and education-**Support conservation research, monitoring, education and information exchange related to local, national and global environmental and conservation issues.
- **Species-oriented projects:** Certain species have been identified as needing a concerted and specifically directed protection effort. Project Tiger, Project Elephant and Project crocodile are examples of focusing on single species through conserving their habitats.

Ex situ Conservation

☆ **Botanical gardens, zoos, etc.** To complement *insitu* conservation efforts, *exsitu* conservation is being undertaken through setting up botanic gardens, zoos, medicinal plant parks, etc by various agencies. The main objectives of this garden are–

- *exsitu* conservation and propagation of important threatened plant species,
- serve as a Centre of Excellence for conservation., research and training,
- build public awareness through education on plant diversity and need for conservation. A number of zoos have been developed in the country.

These zoological parks have been looked upon essentially as centres of education about animal species and recreation. They have also played an important role in the conservation of endangered animal species.

Gene Banks: Ex situ collection and preservation of genetic resources is done through gene banks and seed banks. The National Bureau of Plant Genetic Resources (NBPGR), New Delhi preserves seeds of wild relatives of crop plants as well as cultivated varieties; the National Bureau of Animal Genetic Resources at Karnal, Haryana maintains the genetic material for domesticated animals, and the National Bureau of Fish Genetic Resources, Lucknow for fishes.

Cryopreservation: ("freeze preservation") is particularly useful for conserving vegetative propagated crops. Cryopreservation is the storage of material at ultra low temperature of liquid nitrogen (–196 degree celcius) and essentially involves suspension of all metabolic processes and activities. Cryopreservation has been successfully applied to meristems, zygotic and somatic embryos, pollen, protoplasts cells and suspension cultures of a number of plant species.

Conservation at molecular level (DNA level): In addition to above, germplasm conservation at molecular level is now feasible and attracting attention. Cloned DNA and material having DNA in its native state can all be used for genetic conservation. Furthermore, non-viable material representing valuable genotypes stored in gene banks can all be used as sources of DNA libraries from where a relevant gene or a combination of genes can be recovered.

Legal measures: Market demand for some body parts like bones of tiger, rhino horns, furs, ivory, skins, musk, peacock feathers, etc. results in killing the wild animals. The Wildlife Protection Act (1972) contain provisions for penalties or punishment to prevent poaching and illegal trade. India is also a signatory to the Convention on International Trade in Endangered Species of Wild Fauna and Flora (CITES). The Convention come into force on 1st July, 1975. In addition to this, India is also a signatory to Convention on Biological Diversity (CBD), which it signed on 29th December, 1993 at Rio de Janeiro during the Earth Summit. The Convention has three key objectives:

Conservation of biological diversity,

Sustainable use of biodiversity, and

Fair and equitable sharing of benefits arising out of the utilization of genetic resources.

Questions

Short Answer Type Questions

1. Define species richness.
2. Expand IUCN, CITES.
3. Give the number of National parks in India.
4. Name to *ex-situ* methods of biodiversity conservation

Essay Type Questions

1. Explain various levels of biodiversity.
2. Write down the Importance of biodiversity.
3. What are the causes of depletion of biodiversity?
4. What are the various methods for the conservation of biodiversity?

Index

B

C

R

S

W

X

Z